Wearables in Industrial Hygiene

The primary goal of any industrial hygienist is to prevent hazards before they affect workers. However, safeguards can fail, and human error is inevitable, leading to potential overexposure to occupational hazards. This book explores how to create a dynamic and responsive workplace, emphasizing the importance of early detection and intervention using modern technology through wearable monitors and sensors.

This book explores the integration of wearable technology in industrial hygiene. It examines the basics of wearables, from sensor detection and threshold comparison to real-time data interpretation, and investigates how wearables can bring meaningful improvement to the way industrial hygienists and workers access and utilize information to enhance workplace safety. By offering near-instantaneous feedback, these devices allow workers to make informed decisions and understand the impacts on work procedures for their own health. Real-life scenarios and case studies illustrate the practical applications and benefits of the current and emerging technologies, highlighting how they can transform safety protocols and health outcomes in real time. Readers of this title will develop a thorough understanding needed to navigate the future of industrial hygiene safety through wearable technology.

Wearables in Industrial Hygiene: Improving Safety through Real-Time Health Monitoring is an essential resource for practitioners and engineers in occupational health and safety, industrial hygiene, sensor technology, computer science, and biomedical fields. Its appeal also extends to policymakers and industry leaders aiming to implement the latest safety measures in their organizations.

Wearables in Industrial Hygiene

Improving Safety through Real-Time Health Monitoring

Michelle Stewart and Stephen Lee

CRC Press
Taylor & Francis Group
Boca Raton London New York

CRC Press is an imprint of the
Taylor & Francis Group, an **informa** business

Designed cover image: Hannah Stockstad (painterdog_art)

First edition published 2027
by CRC Press
2385 NW Executive Center Drive, Suite 320, Boca Raton FL 33431

and by CRC Press
4 Park Square, Milton Park, Abingdon, Oxon, OX14 4RN

CRC Press is an imprint of Taylor & Francis Group, LLC

ISBN: 9781032914800 (hbk)
ISBN: 9781032931821 (pbk)
ISBN: 9781003564782 (ebk)

DOI: 10.1201/9781003564782

Typeset in Palatino
by Deanta Global Publishing Services, Chennai, India

Contents

About the Authors

Michelle Stewart is a Certified Industrial Hygienist and Senior Health and Safety Advisor for Chevron, currently residing in Salt Lake City, UT, USA, where she is based in project-based consultation, R&D, and advocates for industry advancement. Her expertise extends to strategic planning, emergency response, and day-to-day health and safety for gas, chemical plants, labs, refineries, alternative/renewable energy technologies, and drilling sites. With 16 years spent at Chevron, she champions a diverse and inclusive workforce and community wellbeing to yield better results. She holds a BSc in Environmental Studies from Sonoma State University, CA, USA, and a Master's in Public Health from the University of California, Berkeley, CA, USA.

Stephen Lee is a Chief Technical Officer and Co-Founder at Epicore Biosystems in Cambridge, MA, USA. Possessing degrees from Dartmouth College and an M.S.E.E. from Northeastern University, Boston, MA, USA, in Electrical Engineering, he has held roles at Analog Devices, Baydin, MC10, and Neurolux. He is Bioelectronics and Biosensors Advisor at the Querrey Simpson Institute for Bioelectronics at Northwestern University, USA.

Acknowledgments

We would like to thank Roozbeh Ghaffari and AJ Aranyosi for their advice and review. To our families for their ongoing support and encouragement. Thanks also extended to Tiye Foley, Laura Moore, Kathryn Underhill, Michael Wasik and Louisa Reid for their collaboration and encouragement. A heartfelt thanks to our colleagues and research partners who have been on this journey with us and will continue to innovate, inspire, and help to create a work environment where "zero" is achievable.

1

History of Sensors and Wearables

Introduction

Detecting contaminants in the environment is often part of a workplace injury and illness prevention plan or program. Understanding work processes and potential exposures is the key role and responsibility of a workplace industrial hygienist, allowing employers to better understand where to design workplace control systems (e.g., engineering controls or personal protective equipment) and procedures to help minimize worker exposure to potential workplace contaminants. Measuring workplace exposures with accuracy and precision has been an evolving science among the public health community (e.g., industrial hygienists, toxicologists, epidemiologists, etc.). This effort faces various limitations that include accepted practices, the public health community's understanding of physical limitations, and access to scalable resources. As detection technologies have evolved, so has our ability to detect contaminants more accurately, which furthers our confidence in results that are deemed not detectable.

Sensors and wearable devices have radically changed how we interact with our environment. With sensors, what we cannot see with the naked eye becomes visible. When used for workplace health and safety applications, sensors highlight the relevance of workplace industrial hygiene. To better appreciate the resources and visibility we have around capturing data, we will explore the history and evolution of sensors and wearables. By examining both the historical context and the technological milestones, we gain deeper insight into the increasing reliance on sensors and wearables in daily life and industrial practice.

Early Development of Sensors

The origins of sensor technology are grounded in basic physical principles and an urgent need to detect environmental dangers before they harm us.

DOI: 10.1201/9781003564782-1

Early detection strategies were strikingly pragmatic. For instance, in the 1800s, coal miners employed canaries as a sentinel species, using them as living detectors of toxic gases such as carbon monoxide and methane. Canaries would manifest symptoms before humans could detect toxic gases with their own senses, thus providing a literal early warning system. The phrase "canary in the coal mine" has since become an idiom for early warning, but its meaning was once quite factual.

Even as technology advanced, the basic principle of finding and measuring interactions with our environment has remained. Through accurate measurement and timely detection of potential workplace contaminants, workers are better equipped to avoid hazardous environments that could potentially impact long-term health. Diligent environmental monitoring has long been the best practice to help safeguard worker health, and the legacy of biological sentinels has persevered in subsequent sensor development.

From Biological Sentinels to Early Sensors

Deploying animals as environmental indicators was eventually supplanted by mechanical and electronic devices. The earliest mechanical sensors aimed to detect physical changes like temperature, pressure, and force. These simple instruments laid the groundwork for the sophisticated systems we utilize today.

Breakthroughs in Sensor Technology

The 19th and early 20th centuries witnessed important discoveries in sensor science. In 1821, Thomas Johann Seebeck invented the thermocouple, marking a pivotal advancement for temperature measurement. When two different metals are joined and exposed to a temperature gradient, an electrical voltage is generated. This allowed for temperature measurement via the conversion of thermal energy to electrical energy. This practice is still fundamental to modern thermocouple sensors.

In 1880, the Curie brothers, Pierre and Jacques, discovered the piezoelectric effect, where certain crystals generate an electric charge when subjected to mechanical stress. This led to the development of piezoelectric sensors, allowing the measurement of pressure, acceleration, and force. These sensors quickly found applications across industrial automation, medical devices, and consumer electronics.

By 1905, Albert Einstein's research into the photoelectric effect led to the creation of the photovoltaic cell, launching the field of light sensing and furthering light absorption and energy harnessing. Sensors in workplace health and safety programs with this technology are now used in light meters, which help detect light in support of workplace compliance and human performance (Colasante et al., 2024) (OSHA, 1982).

In the mid-20th century, instant colorimetric technologies were developed, such as the urine dipstick test by Helen Free (Free et al., 1960). Colorimetric technologies are also widely used in industrial hygiene sample monitoring to quickly detect contaminants. Although using colorimetric tubes to detect contaminants isn't as quantitative as using a pump and sampling tube, this method is a helpful tool during emergencies.

Expansion into Modern Sensing Modalities

As the 20th century progressed, sensor technology diversified and matured. Capacitive sensors, which detect changes in electrical storage, and inductive sensors, which are sensitive to electromagnetic field variations, expand the variety of measurements. These sensors became vital for proximity detection, level measurement, and accurate positioning in manufacturing and automation.

Each development was built upon prior discoveries, with continuous innovation driving enhanced accuracy and reliability. The groundwork laid by these early sensors would enable subsequent leaps in electronic and digital detection, facilitating real-time monitoring and data collection.

The Rise of Direct-Read and Electronic Sensors

One of the most transformative trends has been the shift from laboratory-based analysis to direct-read, field-deployable detection. Early colorimetric sensors, simple tubes, or substrates that change color in the presence of specific chemicals, provided the initial proof of concept for real-time, accessible measurement. Electronic sensors took this a step further, translating chemical or physical changes into digital values displayed instantaneously.

Workplace health, safety, and environmental monitoring remain a core application for sensor technology. Industrial hygienists deploy a spectrum of devices, such as stationary sensors, direct-read instruments, and wearables, to gain a better understanding of potential workplace exposures and where to focus improvements. The evolution from biological sentinels to electronic wearables encapsulates the broader sensor evolution: from reactive detection to proactive prevention. These modern devices provide instant alerts, protecting workers without ethical or practical drawbacks. Likewise, wearable environmental sensors now allow for continuous monitoring of airborne particulates or other aerosols, enabling dynamic risk assessment and exposure reduction in real time.

Today, direct-read devices and sensors focus on environmental media (i.e., air, water, soil), and yet the underlying principles support a broadening range of applications from industrial hygiene to public health surveillance. As recently as 2024, scientists like Tara Hummel, Georgia Colasante, and Keana Makari have expanded the use of sensors for disease detection where a single sample coupled with a sensor will provide a real-time diagnosis for diseases such as tick-borne illnesses (Colasante et al., 2024).

Wearable Sensors: Definitions and Capabilities

The sensors described so far are typically found in stationary fixed sources in the field, in the plant, or on a lab bench. Conversely, wearable sensors are portable, electronic devices equipped with detection elements that measure physical, chemical, or biological parameters within a user's near-field environment. Their proliferation has enabled continuous data collection and personalized feedback, revolutionizing approaches to health, safety, and performance optimization.

The distinction between environmental and physiological sensing is increasingly blurred as wearables integrate multiple modalities. Contemporary devices can monitor ambient temperature, humidity, and air quality (environmental aspects) as well as user-specific metrics such as movement, heart rate, and perspiration (physiological).

Biowearables: Personalized Health Sensing

The emergence of biowearables (wearables + biosensors) marks a shift from environmental detection to directly measuring physiological responses. These biosensors track an individual's unique reaction to environmental or physical exposures, supporting personalized health monitoring and precision medicine. Examples seen in mainstream society include continuous glucose monitors for diabetes management, wearable ECG patches, and emerging platforms for tracking hormonal or metabolic states. In the workplace, biowearables could include sweat sensing for dehydration or sensing for contaminant exposure, core body temperature, or ergonomic strain.

The Advent and Growth of Wearables

The marriage of sensors and wearability marked a technological and cultural turning point. While early sensors resided mostly in industrial or laboratory settings, integrating sensing technology into portable or wearable devices launched a new capability—real-time personal monitoring.

The earliest wearable devices were often experimental or gimmicky, such as a calculator on the wrist or a basic pedometer. As electronic components shrank and wireless communication through Bluetooth® became more prevalent, wearables became more feasible for workplace employee monitoring and public health (Huhn et al., 2022). The digital watch was just beginning;

soon, fitness trackers, smartwatches, and medical monitoring wearables became everyday accessories.

The key was not only miniaturization but also the integration of advanced sensors, such as accelerometers, gyroscopes, optical sensors, and biosensors. These devices enable real-time feedback, providing personalized data and actionable insights for users. The convergence of miniaturization and workplace integration has placed meaningful, actionable information directly into the hands (and onto the bodies) of individuals, empowering the modern worker in their own health advocacy, injury, or illness prevention and mitigation.

Technological Advancements in Sensors and Wearables

The last four decades have witnessed an explosion in sensor and wearable technology. Within my fifteen-year career as an industrial hygienist, I have witnessed the evolution of these advances and uses in the workplace. The development of MEMS (Micro-Electro-Mechanical Systems) introduced tiny, highly accurate sensors for measuring motion, pressure, and other parameters. Wireless networking allows sensors to communicate instantaneously with other devices and with cloud-based analytic platforms (we'll further explore this concept in the chapter discussing Software as a Service or SaaS).

Direct-read sensors, including both colorimetric and electronic media, shifted detection from laboratory analysis to on-the-spot measurement. Advancements in color-changing chemical tubes allowed for not only the detection of hazardous gas in the field, but also more quantification through calibrated industry standards and testing. Electronic sensors further refined this process by capturing, interpreting, and displaying results in digital formats. Such approaches brought real-time insights into the workplace, public spaces, and individual wearers.

Major Milestones in Wearables

The evolution of wearable technology is marked by several distinct milestones:

- The introduction of the first digital watches and calculator watches in the 1970s and 1980s.
- The arrival of heart rate monitors and pedometers for athletes.
- The mass-market breakthrough of fitness trackers and smartwatches in the 2010s.

- The development and USA Food and Drug Administration (FDA) approval of medical-grade wearables for continuous glucose monitoring, arrhythmia detection, and so on.

Many of these milestones were not directly influenced or purposed for the workplace. However, each milestone reflects the merging of sensor miniaturization, wireless communication, energy management, and user-centered design, which are important for workplace adoption of wearable technologies.

Industrial and Occupational Health

Industrial hygienists (IH), environmental health specialists, toxicologists, and safety engineers have long relied on environmental sensors to monitor hazards to help mitigate risks. The ability to measure workplace exposures with accuracy and precision empowers employers to design effective control systems and procedures. The traditional method of exposure sampling utilizes media and an active pump. After sampling, the media is transferred to a laboratory where its contents are removed and run through various spectroscopies to determine the probable concentration of a contaminant in the air during the sampling period. Typical sampling periods include a 15-minute time-weighted average or Short-Term Exposure Limit (STEL), or a longer duration, such as an 8- or 12-hour time-weighted average. Within that sampling period, there may have been episodic exposures or peaks that were undetectable by this sampling technique. Direct-read instruments, therefore, allow an industrial hygienist to see, sometimes in real time, those seemingly undetectable peaks. IHs can now pinpoint exactly when a potential exposure occurred and follow up with the worker afterward—either at the end of the day or following an alert. What could have potentially been missed as an improvement opportunity is now accessible, advancing occupational health from being reactive to proactive.

The role of the industrial hygienist has likewise evolved as technology has improved, requiring not only an understanding of physics and detection technology but also skill in interpreting and communicating complex data. Workplace injury and illness prevention, along with industrial hygiene, will become timelier, efficient, and effective through the adoption of these advanced sensors.

Healthcare and Personal Monitoring

In healthcare, wearables and biosensors have opened new frontiers in personalized medicine. Devices can now monitor heart rate, respiration, blood pressure, glucose levels, and even biochemical markers, delivering real-time feedback to both users and clinicians. Such technology supports early intervention, chronic disease management, and even predictive analytics for health risk assessments. Industrial hygienists monitor and detect probable exposures, and when combined with occupational health doctors and clinicians who monitor health outcomes, a more accurate picture of worker health emerges. Biowearables lay the groundwork for these two professions to work even more closely while utilizing the same data.

Sports, Fitness, and Everyday Life

Wearables have become ubiquitous in sports and fitness, tracking performance, movement, and physiological responses in real time. These insights enable athletes to optimize training, prevent injury, and push the boundaries of human capability. For everyday users, wearables promote engagement in personal health and wellness by providing easy access to actionable personalized data. When considering the industrial population or industrial athletes (Sevier Tl, 2000), these sensors are adaptable to the workplace ecosystem and can assist in monitoring workers' boundaries. These sensors could detect and help prevent injury and illness.

Environmental and Public Health Monitoring

Beyond individual health, sensors play a critical role in environmental monitoring through detecting air and water quality, radiation, and other hazards. The deployment of large networks of distributed sensors enables public health agencies to respond rapidly to emerging threats and identify patterns that would otherwise remain hidden. Deploying sensors in an industrial plant environment or within a community is now a widespread practice. Industrial hygienists often use environmental monitoring to focus on workplace sampling campaigns. The advancement opportunity in this space is to connect environmental and wearable sensors providing a bigger picture and for easier analysis and interpretation.

Challenges and Limitations: The Physics and Reality of Sensing

Despite extraordinary progress, advancing sensor technology remains constrained by fundamental physics and practical limitations. Sensitivity, selectivity, power consumption, and scalability are persistent challenges. Even as use cases expand, not all desired capabilities are achievable given current materials, methods, and energy constraints.

For instance, the miniaturization and portability of sensors often come at the expense of battery life and sustained accuracy. Need and demand persist, and as technology evolves in the areas of weight, energy, and real-time processing, so will innovation.

The Nano Revolution

Nanotechnology has promised to revolutionize sensor research and development by unlocking new capabilities via quantum-scale materials and phenomena. While advances have occurred (Wang & Fu, 2024), particularly in sensitivity and surface chemistry, many anticipated breakthroughs have been hindered by manufacturing complexity, integration issues, trust in technology, and power management challenges. The bottleneck remains: batteries, electronics, and infrastructure (such as hard-wiring and wireless communication) have yet to fully catch up to the theoretical possibilities envisioned by nano researchers. True disruption in portable sensor detection is yet to be seen, particularly in miniaturized and sustainable power sources. However, nanotechnology is being tapped to solve this limitation (Wang et al., 2022).

Regardless of the bottleneck, improving sensor sensitivity and selectivity with nanotechnology will, at some point, converge with breakthroughs in power, leading to significant advances through automated workplace hazard characterization and response.

Power, Miniaturization, and the Portable Sensor Paradox

The push for smaller, lighter, and more integrated sensors is continually undermined by energy storage and delivery constraints. Wearable devices must balance form factor and battery life, with miniaturization often necessitating innovative energy-harvesting methods, ultra-low-power electronics, and careful system design. Progress continues, but the quest for an ideal portable sensor that is also affordable, reliable, accurate, and unobtrusive remains at the tips of inventors' fingers. For instance, industrial hygienists want to see measurement integrations averaged over each minute. This is 60 measurements for one minute of data. Those measurements require a certain amount of energy consumption or battery life, and therefore, the amount of time that the device may be used is finite. In an industrial setting, 8 to 10 hours of measurement are ideal, so the amount of data obtained must be balanced with the length of time a wearable must be worn, not only for a single day but for a work week.

Looking Ahead: The Future of Sensors and Wearables

As technology continues to evolve, the frontier of sensors and wearables promises even greater integration, accuracy, and benefit to society and the workplace. Advancements in artificial intelligence and machine learning will enable predictive analytics, anomaly detection, and personalized recommendations to further empower the worker. The eventual convergence of nano, bio, and info technologies may yield truly autonomous, self-powered, and multifunctional sensing platforms that would help provide more powerful and timely information to support workplace injury and illness prevention programs.

Regardless of the technologies that are created and their advancements, power management, data privacy, interoperability, and cost will continue to shape the direction and impact of future innovation. The goal is clear: to empower individuals, safeguard workers, and inform societies through ever-refined detection and feedback mechanisms.

By understanding the evolution of sensors and wearables, we are better poised to shape their future and harness their potential for a safer, healthier, and smarter world.

2

Sensors and Wearable Technology

Introduction

Sensors and wearable technology have revolutionized the way we perceive and manage our surroundings and personal well-being. This comprehensive discussion delves into the fundamental principles of sensors, their several types, detection mechanisms, and the rapidly evolving landscape of wearable devices. We examine applications across industries, compare technological paradigms, and explore the challenges and future trends shaping this dynamic field.

Defining Sensors

At the most basic level, a sensor is a device that responds to physical stimulus, such as heat, light, sound, pressure, magnetism, or movement, and transmits a resulting impulse for measurement or operating controls. In industrial hygiene and numerous other domains, sensors serve as essential tools, enabling real-time detection and mitigation of hazards. Their design accounts for factors such as sensitivity, environmental interactions (e.g., with water vapor), shelf life, and the overall precision and accuracy required for extended usability.

The Role of Sensors in Modern Society

The proliferation of sensors underpins many facets of modern life, from ensuring safety in industrial environments to enabling personal health monitoring. Real-time detection and decision-making capabilities drive

DOI: 10.1201/9781003564782-2

innovation and push the boundaries of what sensors can achieve, fueling a commercial market that spans gas detection, environmental monitoring, healthcare, transportation, and more.

Types of Sensors and Detection Principles

Gas Detection Sensors

Gas detection remains one of the most vital applications for sensors, particularly in industrial safety. Figure 2.1 illustrates the primary types which include catalytic sensors, electrochemical sensors, photoionization detectors (PIDs), infrared sensors, and metal oxide semiconductor sensors (ISC, 2025)

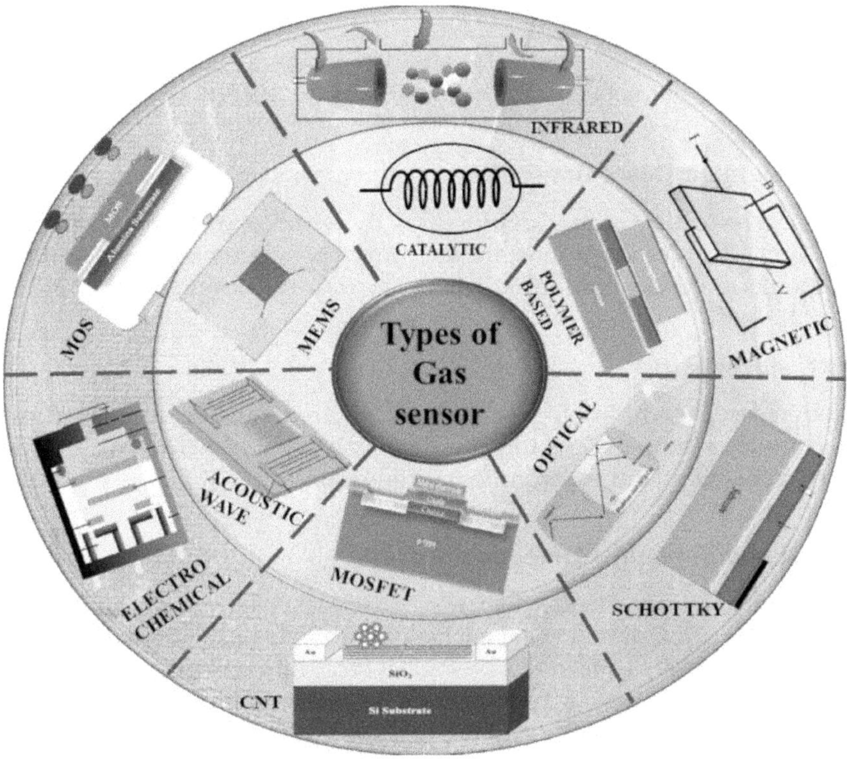

FIGURE 2.1
Types of gas sensors. Note: Illustrates various gas sensor technologies, such as catalytic sensors, electrochemical sensors, photoionization detectors, infrared sensors, and metal oxide semiconductor sensors. The image compares their detection principles and highlights their applications in industrial safety and environmental monitoring (Sagarika et al., 2024) (image used with permission).

(Sagarika et al., 2024). Each offers a unique mechanism for detecting different classes of gases and vapors.

Catalytic Sensors

Catalytic sensors, also known as catalytic bead sensors, are widely utilized for detecting combustible gases. They function by oxidizing gases in a catalyst chamber, where the resultant temperature change is converted into an electrical signal. Typically, these sensors are calibrated to detect concentrations as a percentage of the lower explosive limit (LEL) of gases such as methane and propane.

Electrochemical Sensors

Electrochemical sensors are designed to detect specific toxic gases such as carbon monoxide (CO) and hydrogen sulfide (H_2S). They are comprised of electrodes immersed in an electrolyte; when a target gas contacts the working electrode, an electrochemical reaction generates a current or voltage proportional to the gas concentration (Sagarika et al., 2024).

Photoionization Detectors (PIDs)

PIDs target volatile organic compounds by ionizing molecules with ultraviolet light. The resultant ions are measured, providing sensitive and rapid detection of even low concentrations of hazardous compounds.

Other Sensor Types

Additional sensor technologies, such as infrared sensors and metal oxide semiconductor sensors, expand the detection capabilities across a broader spectrum of gases and environmental conditions.

Physical and Biological Detection Sensors

Sensors are not limited to gas detection. Advances in biosensors have enabled real-time detection of biological agents in both field and laboratory settings. Biosensors typically consist of a biological recognition element (such as antibodies or DNA probes), a transducer that converts biological interactions into measurable signals, and a data processing system (Zhao et al., 2014).

General-Purpose Sensors

- Proximity Sensors: Detect the presence of nearby objects without physical contact (Robocraze, 2022).
- Hall Effect Sensors: Measure magnetic field strength.

- Load Cells: Convert force into electrical signals.
- Accelerometers: Measure acceleration and motion forces.
- Temperature Sensors: Monitor thermal variations.
- Ultrasonic Sensors: Measure distance using sound waves.
- Light Sensors: Gauge light intensity within environments.
- PIR Motion Detectors: Detect movement based on infrared radiation.

Comparative Analysis of Sensor Technologies

Catalytic vs. Electrochemical Sensors

A comprehensive comparison reveals that while catalytic sensors are favored for broader detection of combustible gases, electrochemical sensors offer higher selectivity and sensitivity for specific toxic substances. Catalytic sensors feature fast response times but require frequent calibration. Conversely, electrochemical variants benefit from longer lifespans but may respond more slowly due to the diffusion process.

Evolution of Detection Mechanisms

Detection mechanisms influence operational efficiency, accuracy, and application context. For instance, catalytic sensors utilize temperature changes from gas oxidation, while electrochemical sensors rely on redox reactions at the electrode interface. The benefit of detection mechanisms is that they can be thought of as self-sustaining technology, allowing a longer sensor life and aiding in the compaction of the sensor. The choice between mechanisms is dictated by the target substance, environmental context, and required detection limits.

Real-Time and Near Real-Time Detection

Real-Time Sensor Detection

Real-time sensor detection is characterized by immediate acquisition, processing, and communication of data, with latencies measured in milliseconds. This immediacy is vital in critical environments where swift intervention can prevent accidents or mitigate risks, such as in gas leakage scenarios, where alarms must be instantaneous.

Near Real-Time Sensing

Near real-time detection introduces slight delays, ranging from seconds, minutes, or even hours, due to power-saving techniques, processing, or transmission lags. While not instantaneous, near real-time systems are invaluable for trend analysis and monitoring scenarios where some delays are tolerable.

Combination of Both

Many workplace monitoring devices include a mix of real-time sensing paired with near-real-time data transmission. The detection and physical device alarming could be instantaneous; however, the data may require connecting to a device within Bluetooth® range or accessing a cellular or Wi-Fi network.

Conclusion

Sensors and wearable technology form the backbone of modern environmental, health, and safety monitoring programs. Their evolution has been shaped by advances in materials science, electronics, communications, use cases, and data analytics. As challenges are addressed and new opportunities emerge, these technologies will continue redefining how we interact with our environments and our own bodies, driving progress across countless domains.

3

Industrial Hygiene Opportunities

Case Study: Welding in a Confined Space

I remember when direct-read sensors were first small enough to be worn comfortably by a worker. This was not very long ago; it was the early 2000s when I used my first direct-read sensor for an exposure monitoring sample of carbon monoxide while a worker welded in a confined space. We were executing a sampling strategy to understand if the ventilation we designed and implemented was effective in removing welding fumes and carbon monoxide from the welder's breathing zone. The long turnaround time to obtain results for a welding fume sample (even when rushed) meant the worker would continue wearing a self-contained breathing apparatus until the data established any potential exposure to metals would be mitigated by wearing a half-face respirator, or perhaps no respirator if supported by the data. Welding in a confined space at times includes wearing layers of protective clothing and sweating in your full-faced mask, while your welding hood potentially shifts or needs tightening as you perform the perfect weld to pass QA/QC. In these circumstances, workers are keen to obtain welding fume sample results. Similarly, readily available data allows employers to determine whether both a personal air system and a ventilation system are needed for the entirety of a job. I recall a maintenance event manager at one point asking if we could physically bring the lab to our job site to expedite the process. However, there wasn't adequate time to work through that solution. Our carbon monoxide sensor provided real-time data, and our metal concentration lab results that were available 48 hours later (after interpretation by a Certified Industrial Hygienist) generally established that our generalized and localized exhaust ventilation (LEV) was effective. We typically would not know an individual worker's potential metal exposure results until the maintenance event ended, and we compared the worker's pre-and post-blood level metal concentrations. What we observed during this study, and one observation that fueled this author's curiosity, was whenever the personal carbon monoxide monitors alarmed, workers would adjust their LEV, their monitor alarms would stop, and workers would continue working

DOI: 10.1201/9781003564782-3

without any significant interruption. We could also predict which activity would lead to a potentially higher concentration of metal fumes in a worker's breathing zone based on how tasks were performed in the confined space (e.g., grinding vs. TIG welding) and we further noticed a difference between years of experience and exposure: some professional laborers and welders had been doing this trade for years and were flawless in their execution, resulting in hardly any emissions. Their blood-level metal concentrations also reflected this level of experience.

An Introduction to Industrial Hygiene Concepts

To understand the importance of knowing the concentrations of an aerosol or vapor is to understand certain key industrial hygiene concepts. Simply put, Occupational Exposure Levels (OELs) are the concentrations at which a worker may be exposed to a certain contaminant until an intervention must be introduced. An intervention or mitigation may include stopping any further exposure by removing the hazard to begin with (elimination), removing the person for periods of time from an environment (administrative control), donning personal protective equipment (PPE), or modifying equipment (engineering controls), etc. These OELs are agreed upon by a body of professional experts, inclusive of toxicologists, epidemiologists and occupational health physicians, who review various laboratory capabilities, clinical case studies, and epidemiological studies to determine the acceptable limits of exposure to a worker before a health endpoint is observed. To determine potential exposures in a workplace, a sampling campaign is designed and executed based on work activities and hazard characterization. These sampling campaigns may require years of data collection to build a statistically significant exposure profile around the worker. Let's walk through the process of collecting a single worker sample to help quantify the magnitude of time and effort necessary to conduct these campaigns using traditional analytical sampling methods. After a qualitative review of procedures, observations/shadowing, and safety data sheets, an IH will select the pump (ensure its calibration is up to date by the manufacturer and charged), ensure sampling media is not expired, perform a calibration of the sampling media, document sample identification, and find the pertinent worker. Once with the worker, the IH would confirm the activities the worker is planning to perform that day to ensure the sample being taken is appropriate. The IH would then gather worker identification details for reporting, don the sampling assembly on the worker, provide instructions, turn on the pump, and validate its functioning as expected. For a full-shift sample, the IH would return either 8, 10, or 12 hours later to interview the worker on their activities that day, when the worker took breaks, and whether there were any unusual sounds, odors, etc. Following the worker interview, the IH would return to the lab, secure the used sampling media, recalibrate the pump, document results, document the chain of custody (CoC), create a shipping label, package the media and CoC, and mail the sample. Once the sample arrives at the quantitative laboratory, an IH would, on average, wait one week to receive the results. Finally, the IH would calculate the results based on actual sampling time to generate the time-weighted average, compare those results against the OEL, develop findings and a report, and send those findings and reports to the worker. Phew. Now imagine the improved effectiveness of an industrial hygiene program if all workers wore digital wearables that would alert the industrial hygienist whenever a concentration exceeded a pre-set threshold; there potentially would be reduced need for lab analysis while experiencing timely and targeted conversations with the worker.

Leveraging Wearables for IH Data Management and Collection

Industrial Hygiene qualitative and quantitative exposure assessments include a lot of data collection and management. When performing monitoring on workers, it is important that the IH builds a good trusting relationship with the worker; otherwise, the information they receive may be limited or constrained. In those small moments with the worker, the IH is often restricted to obtaining only the information that is critical for reporting. However, more data to fully paint the conditions of a work environment during a sampling event is invaluable. When utilizing digital wearables, the devices can easily gather information for the IH, including GPS identification of location, date, time, and duration of a potential exposure, allowing the IH to trace back the history of equipment operation and/or immediately contact employees to learn what they were doing during that detection event. When thinking about appropriate data to collect for IH digital wearables, there are various industry standards or best practices, such as the American Industrial Hygiene Association (AIHA) recommended data collection list (Figure 3.1) Documenting data is important to understand the statistical significance of exposure as well as the overall effectiveness and defensibility of an IH program.

> However, many of these data sources are lacking contextual information that could allow them to serve even greater purposes. For example, the OSHA IMIS records do not include information on the sampling method, sampling instrument, or site or work area.
>
> **(Shockey et al., 2020)**

Where real-time detection and wearables can help is with standardizing data collection, tagging data from other data sources, pinpointing worker

	Data Group	Summary Contents
A.	Facility/Site Information	Basic descriptive information regarding the facility or site.
B.	Survey Tracking Information	Summary survey identification and tracking information.
C.	Work Area Information	Information about the area(s) where the exposure measurements/assessments were performed (location, type, ambient conditions).
D.	Employee Information	Information about the employee(s) whose exposure was assessed/measured, nature of job, exposure groups.
E.	Process and Operation Information	Information about process(es), operation(s), exposure source(s).
F.	Chemical Agent Information	Information about the chemical agent(s) monitored.
G.	Exposure Modifier Information	Information about modifiers of exposure (representativeness, pattern, frequency, work load).
H.	Sample Information	Information about the sample(s) (reason, date, duration, type).
I.	Sampling Device Information	Information about the sampling device used.
J.	Engineering Controls Information	Information about engineering controls (type, condition, usage)
K.	Personal Protective Equipment Information	Information about respirators; gloves; protective clothing; hearing, eye, face, and foot protection.
L.	Chemical Exposure Results	Concentration measured, TWA estimates, limit of detection.
M.	Noise Exposure Results	Noise exposure estimates (frequencies, intensities, intermittences).

FIGURE 3.1
Required data for industrial hygiene monitoring. ***Source***: Lippmann & Gomez, 1996. © 1996 JOEH, LLC, reprinted by permission ofInforma UK Limited, trading Taylor & Francis Group,https://www.tandfonline.com on behalf of JOEH, LLC.

tasks that may have opportunities for improvement, and automating data analysis, reporting, and identifying opportunities. Long-term, with an implemented feedback mechanism, workers can receive instantaneous data from wearables while maintaining confidence in their safety. The time that was spent by the IH calibrating pumps or donning equipment on the worker can now be spent on other program activities.

Case Study: Sweat-Sensing Biowearable

What sold me on biowearables was the insight gained from the data from a sweat-sensing wearable. The biowearable demonstrated greater accuracy for individual workers compared to environmental measurements or environmental-only wearables paired with antiquated calculations that are based on a specific subset of workers and data from the 1970s and 1980s, even after applying correction factors for work intensity, personal protective equipment, and related variables. What this technology did for the profession is measure sweat as a surrogate to the heat strain workers experienced due to exposure to confounding factors, rather than estimate heat strain using an algorithm that considers estimated confounding factors (Rothfusz, 1990). The traditional industrial hygiene approach would be to estimate or directly read the confounding factors and plug them into a calculation based on the average "standard man" of 70 kg (154 lb.) body weight and 1.8 m^2 (19.4 ft^2) body surface. This may lead to an over- or underestimation if the worker doesn't meet the "standard man" criteria. Scientists argue this inaccuracy doesn't matter, as the clinical signs and symptoms of heat stress are the same, no matter the differences in the body. However, to identify heat strain before it becomes clinical (evolving into heat stress) allows interventions (Ayotte & Corcoran, 2018) to be more effective and targeted in keeping a worker healthy and able to work, rather than just reducing a potential injury to the worker.

When deploying or testing technology, it can start from a proof of concept and a small data set to help form a hypothesis on which to design future testing. For biowearables, some hypotheses to consider may be monitoring physiological results and predicting risk, real-time vs. delayed results, and risk or behavior change through direct feedback to the user to prevent disease.

Analysis of the data from a sweat-sensing wearable demonstrated that the interventions recommended by local environmental conditions and using the Heat Index were not as mitigative as what the patch was recommending for the worker (Stewart et al., 2023). Taking it a step further, when comparing a population-based threshold (Heat Index) to personalized results from the use of biowearables, the risk level went from "Caution" to "Extreme Caution" or even "Danger." With a small sample size, there isn't enough data to determine actual risk; however, it does provide enough evidence to continue to explore a more personalized approach to mitigate heat strain as part of a workplace Heat Illness Prevention Program (Stewart et al., 2023).

Conclusion

The American Industrial Hygiene Association (AIHA), National Institute for Occupational Safety and Health (NIOSH), and American Conference of Governmental Industrial Hygienists (ACGIH) have been monitoring the progress and helping to define use cases for sensors and wearables to support the practice of industrial hygiene. Technical frameworks and thorough bodies of knowledge have been and continue to be developed around direct-reading instruments, sensors, big data, and artificial intelligence (AI), among other areas (AIHA, 2016) (Cauda et al., 2022). Continued advocacy and education will further shape problem statements, collaborative research, and integrating technology into the workplace. The difference between population health thresholds versus personalized thresholds and measuring them through non-invasive technology is the disruption we've been waiting for.

4

Health and Safety Program Improvements

Introduction

In industrial and occupational settings, the effectiveness of health and safety management systems hinges not just on robust procedures and competent persons but also on timely, accurate data about potential and actual workplace conditions and worker exposures. The traditional sequence of risk assessment, hazard analysis, job safety analysis, safe execution of work, verification and validation, and lessons learned provides a structured approach to hazard mitigation and continuous improvement. However, these processes often rely on historical data, delayed feedback, and periodic manual sampling, all of which can limit the system's responsiveness and capacity for real-time intervention. The emergence of wearable devices and real-time sensors is transforming this landscape (Aksut et al., 2024), offering the potential to enhance each step in the safety management system with continuous, personalized, and actionable data. This chapter explores, in depth, how these technological tools can support and elevate each stage of safety management, cultivating a culture of proactive risk management and adaptive safety practices.

Risk Assessment: Data-Driven Hazard Identification

Risk assessment is the foundational step in any safety management system. It involves identifying hazards, evaluating the likelihood of exposure, and estimating potential impacts. Traditionally, risk assessment depends on historical records, static measurements, or sporadic environmental sampling, which may not fully capture the inherent variability or evolving nature of dynamic work environments.

These activities are essential for any health and safety management system to perform work and maintain the health and safety of the workforce.

DOI: 10.1201/9781003564782-4

However, feedback on how well this system is working can often be random, delayed, and time-intensive. The same system can be applied to an industrial hygiene program. Let's take an example of a work activity that requires confined space entry to demonstrate the potential opportunities for workplace wearables.

To start, a risk assessment is performed. In an industrial hygienist's understanding, exposure = receptor × hazard. For the workplace, the receptor is the worker; in our example, let's say this worker is a welder who will be welding stainless steel in a confined space. We know from other studies and various hot work manipulations of stainless steel that hexavalent chromium may elute during welding and that a confined space typically has poor airflow as well as a risk of contaminants staying in the space. Hexavalent chromium is considered a hazardous material and carcinogen with acute effects. To make this scenario more interesting, the type of welding performed on the stainless steel is stick welding, or Shield Metal Arc Welding (SMAW), presenting a higher probability that hexavalent chromium particles of respirable size will be present in the welding fumes.

Wearable devices and real-time sensors introduce a change in thinking. Modern wearables equipped with environmental sensors (e.g., gas detectors, particulate counters, temperature, and humidity sensors) can continuously monitor conditions in and around the worker. This real-time data provides a granular view of exposures as they fluctuate due to changes in process, weather, or worker movement. For example, personal gas detectors can measure the concentration of hazardous substances such as carbon monoxide, hydrogen sulfide, or volatile organic compounds, alerting both the workers and safety managers when thresholds are approached or exceeded. Wearables can also monitor physiological parameters—heart rate, body temperature, exertion levels—that may indicate early signs of heat stress, fatigue, or medical distress.

By integrating this continuous data stream into the risk assessment process, organizations can better identify emerging hazards, quantify real-time risk, and prioritize interventions. Furthermore, aggregating data from multiple workers and locations enables predictive analytics, helping safety professionals anticipate and address risks before they escalate.

Hazard Analysis: Real-Time Characterization and Mapping

Hazard analysis builds upon risk assessment by examining the specifics of each hazard's characteristics, source, and mechanism of harm. In the past, detailed hazard mapping required labor-intensive sampling campaigns and laboratory analysis, often resulting in delayed or incomplete information.

With this understanding of potential hazards, we proceed to analyze how these risks manifest during specific job tasks. When conducting a job hazard analysis for welding on stainless steel in a confined space, a common mitigation against the hazard would be respiratory protection for the worker. Without having data regarding this specific work, we would err on the side of caution and expect that concentration in a confined space, even with general ventilation, would require a full-face in-line supplied air or a full-face air-purifying respirator to provide the best respiratory protection to the welder. Now, by a show of hands, who has worn a full-face in-line supplied air or full-face air-purifying respirator with a welding hood? I'm going to guess the response is close to zero. The configuration is awkward, the visibility is limited, and the worker has a high chance of wearing it improperly or shifting over time, plus the additional overhead costs for the air-tank watch and cartridge replacements. Also, air-filtering respirators don't filter out the carbon monoxide (CO) that may also be generated.

What does the industrial hygienist do? They get out their pumps and cartridges and sample the welder and any other occupants in the space. They plan for and validate the installation and use of localized exhaust ventilation (LEV), in addition to general ventilation, that must be manipulated by the welder or an assistant to ensure it's capturing the CO and hex chrome.

Wearable sensors now allow for dynamic hazard mapping as the work unfolds (Xin et al., 2017). For example, direct-reading instruments integrated into wearable platforms can log data on airborne particulates, toxic gases, or noise levels, correlating exposures with specific activities, locations, or times of day. When workers traverse different areas of a facility or construction site, GPS-enabled wearables can geolocate exposure data, creating exposure "heat maps" that visually represent where hazards are most intense or frequent. This detailed mapping informs targeted engineering controls, ventilation strategies, or administrative interventions. Additionally, some wearables can interface with other smart infrastructures, such as stationary sensors or building management systems, to compare personal exposure data with ambient measurements. Such integration helps distinguish between individual behaviors and systemic environmental risks, refining the precision of hazard analysis.

Job Safety Analysis (JSA): Proactive Hazard Control

Job Safety Analysis (JSA) systematically breaks down each task in a work process, identifies potential hazards at each step, and prescribes control measures. Traditionally, JSAs are updated infrequently and rely on previous experience with hazards that may not account for real-time changes in the work environment.

Alternatively, using the welding scenario, we could have a wearable device on the worker that provides real-time information to the worker and/or the medical clinician about their actual biological exposure to hexavalent chromium. Wearable sensors could continuously monitor air quality and alert workers in real time, potentially reducing reliance on manual sampling and enabling immediate corrective actions. Continuing with our hex chrome scenario, the bioindicator could be a real-time measurement of the actual exposure (i.e., blood levels of hexavalent chromium metabolites), which isn't ideal for the industrial hygienist who aims to prevent exposure. If there were a way to correlate carbon monoxide exposure to hexavalent chromium exposure, or if another early indicator could be detected in real time on the welder and provide the worker with an alert to modify their behavior or technique before they are considered overexposed, that would be ideal.

Wearable devices introduce the possibility of adaptive JSAs, think of dynamic safety protocols that update based on current conditions. For example, as workers don wearables at the start of a job, baseline environmental and physiological data can be captured. As the work progresses, the system can monitor deviations indicating increased risk, such as rising gas concentrations, excessive noise, or elevated stress markers. When thresholds are crossed, wearables can trigger alerts for job stoppage, require additional PPE, or prompt a supervisor review. This creates a feedback loop in which JSAs do not remain static documents; instead, they evolve in response to live data, enhancing the effectiveness of controls.

Moreover, wearable-enabled checklists and task prompts can guide workers step-by-step, ensuring compliance with procedures and immediate acknowledgment of identified hazards. Digital records from these workflows can be reviewed and audited, providing traceable evidence of risk management actions taken in the field.

Execute Work: Enabling Real-Time Protection and Adaptation

The core of any safety management system is the safe execution of work. This is where protective or preventive and mitigative measures are put to the test and where the consequences of gaps in risk management are realized. Wearable devices and real-time sensors directly support workers during this phase by providing immediate feedback and enabling prompt adaptation, empowering the worker by giving them the data and mitigations needed to immediately protect their health and safety.

This leads us to consider how the integration of advanced wearable technology could transform risk management in confined spaces, shifting from reactive to proactive safety practices. If such devices could interface with ventilation systems, automatically adjusting airflow based on detected contaminant levels, or even communicate with external teams for rapid response, they would not only enhance personal protection but also streamline incident data collection and reporting. The ability to aggregate and analyze this real-time exposure data across multiple worksites could guide the refinement of safety protocols, inform targeted training, and foster a culture of continuous improvement in industrial hygiene. As technology evolves, the challenge for health and safety professionals is to ensure these tools are user-friendly, reliable, and robust enough to function reliably in harsh environments, while also maintaining the trust (data privacy) and engagement of workers whose compliance is critical for success.

Back to our welder working in a confined space, now equipped with wearable multi-gas detectors providing instant alerts if hazardous substances like hexavalent chromium or carbon monoxide exceed safe levels. Some advanced wearables can be networked to automatically adjust ventilation systems or activate alarms, reducing the burden on the worker to recognize and respond to invisible dangers. Proximity sensors can warn workers when they approach unsafe zones or enter areas with restricted access. Motion and posture sensors can detect slips, falls, or prolonged periods of inactivity, or trigger a rapid emergency response if needed.

Importantly, wearables can monitor both environmental and personal metrics, enabling a holistic view of safety. For example, a worker's heart rate and core temperature can be tracked alongside airborne contaminant levels, ensuring protective measures address both external and internal risk factors.

Verification and Validation: Data-Driven Confirmation

Verification ensures prescribed controls and safety measures are implemented as intended; validation confirms these measures are effective in reducing risk. Traditionally, this involves post-task inspections, spot checks, and review of incident reports, which are processes that are inherently reactive, delayed, and may miss transient hazards.

Wearable devices and real-time (or near real-time) sensors support continuous verification and validation. By logging environmental and physiological data throughout the work process, these devices provide objective evidence of conditions workers experience. For example, records showing air quality remained within safe limits, or that workers' vital signs stayed in acceptable ranges, can confirm the adequacy of control measures. Conversely, data showing repeated exceedances can prompt immediate remedial action and root cause investigation. Automated reporting functions within wearable systems can simplify compliance documentation, reduce administrative burden, and make it easier to demonstrate regulatory adherence during audits. Integrated dashboards can visualize trends, compare outcomes across teams or shifts, and identify opportunities for further improvement.

Lessons Learned: Continuous Improvement and Knowledge Sharing

The final stage of the safety management cycle is capturing and applying lessons learned. This step relies on accurate, comprehensive data about exposures, incidents, and near misses, as well as on effective mechanisms for knowledge sharing and procedure or process refinement.

Looking ahead, the adoption of wearable technology in industrial hygiene presents an opportunity not just to enhance detection but to personalize safety interventions and foster adaptive responses to evolving hazards. As these systems become more sophisticated, integrating machine learning (Masson, 2024) to anticipate risk patterns and environmental changes could empower both workers and supervisors with predictive insights, eventually turning every shift into a dynamic exercise in risk mitigation. The potential to embed feedback loops directly into work practices, where data-driven alerts prompt real-time or near real-time corrective actions, promises to elevate safety from a procedural compliance exercise to a shared, interactive responsibility. As the boundaries between human activity and intelligent monitoring blur, the success of such innovations will depend on their seamless integration into daily workflows and the creation of a safety culture where continuous improvement is driven not only by incident reports but by proactive, real-time engagement at every level.

Wearable technology dramatically expands the scope and quality of lessons learned. Instead of relying solely on incident reports or worker recollection, organizations can analyze large datasets collected by wearables to uncover patterns, correlations, and emerging risks (Gualandi et al., 2021). Advanced analytics, such as machine learning, can identify predictive indicators of hazardous conditions or precursors to incidents, enabling proactive interventions.

Furthermore, anonymized and aggregated or grouped data can be shared across worksites, industries, or even at the regulatory level, supporting the creation of evidence-based best practices. Digital records generated by wearable devices can be used for training, scenario planning, and simulation exercises, helping to embed a culture of continuous learning and safety innovation.

Challenges and Considerations

While the benefits of integrating wearable devices and real-time sensors into safety management systems are substantial, several challenges must be addressed for successful adoption. In addition to the validation of accuracy and precision of the sensor, user acceptance is paramount; devices must be comfortable, unobtrusive, and easy to operate. Data privacy and security must be safeguarded, especially when personal health information is collected. Robustness and reliability in harsh industrial environments are also critical, as device failure could compromise both safety and trust.

Moreover, the sheer volume of data generated by continuous monitoring requires effective systems for storage, analysis, and interpretation. Investments in infrastructure and training may be needed to ensure organizations can derive actionable information from wearable-enabled safety programs.

Conclusion

The integration of wearable devices and real-time sensors into key safety management system tools has the potential to transform workplace safety from a reactive, compliance-driven obligation into a dynamic, data-informed partnership between workers and technology. By supporting each step within a health and safety program, from risk assessment to lessons learned, paired with objective, personalized, and continuous data, these innovations could empower organizations to identify hazards sooner, intervene more effectively, and foster a culture of continuous improvement. As wearable technology matures and becomes more widely adopted, its potential to save lives, prevent injuries, and optimize health outcomes in the workplace will only grow, potentially making it an indispensable component of modern safety management systems.

5

Evaluating Technology: Lab Bench

Introduction

Sensors are not simply created in a laboratory and subsequently launched into the field. Rather, the physical sensor, as designed, undergoes a series of standardized tests to demonstrate how well the sensor performs against a "gold standard." This level of testing demonstrates how close the sensor's results are to the "real world." The information obtained through lab bench studies provides researchers, regulators, as well as health and safety professionals with a level of confidence in the sensor's results. Understanding a technical specification sheet's values and what that data means for a device in the field is crucial to workers' adoption of technology.

Measurements of Industry Standards—Technical Specifications

An industry standard of measurement refers to a set of established criteria that is widely accepted within a particular field to ensure consistency and reliability in evaluating technologies. In the context of sensors, these standards help determine whether a device meets specific performance requirements, such as accuracy, precision, and detection limits, ensuring the measurements taken are comparable across different applications and environments. Adhering to industry standards is helpful for validating sensor effectiveness and building confidence among users and regulators.

Below are commonly tested specifications often provided to an industrial hygienist on a technical sheet, allowing the practitioner to assess their confidence in a sensor and its housing. Additionally, the AIHA has provided a template for common specifications a practitioner would expect to see from

DOI: 10.1201/9781003564782-5

electronic real-time gas and vapor detection equipment manufacturers (see Figure 5.1) (AIHA, 2022).

Technical Specification Criteria

When reviewing sensor technology, many are tested against standard criteria. Below are brief descriptions of common technical specification criteria.

The limit of quantification (LOQ) and *limit of detection (LOD)* are often used to understand the reliability of a result compared to the true value in its sampling environment. Understanding the limitations of the measuring device, or the detection limit (the lower and upper thresholds of the sensor), allows the industrial hygienist to understand how much they can trust the results compared to their working knowledge of the sensor and conditions during the sampling event. The mean of the blank is compared against the mean of the LOD normal distribution curve and the LOQ normal distribution curve. Results that fall between the blank and LOD mean are often referred to as "noise" and likely are not a true value. The lower the detection range, the higher the sensitivity of the sensor. Results that fall between the blank and LOQ mean are an indication that the contaminant is present and is too low to accurately quantify the value, and is also considered as trace or "noise."

Figure 5.2 illustrates the concept of detection limit and quantification limit by showing the theoretical normal distributions associated with blank,

Sensor Sheet

Manufacturer: ______________________

Instrument Name/Model No.: ______________________

General Information	
Hazard Type	
Sensor Type	
Warranty	

Sensor Performance	
Measuring Range	
Minimum Detection Limit	
Resolution/Sensitivity	
Accuracy/Uncertainty	
Linearity	
Recovery Time	
Response Time T_{90}	
Response Time T_{50}	
Start-up to T90 after 24-hour Storage	
Start-up to T90 after 7-day Storage	
Start-up to T90 after 30-day Storage	
Life Expectancy of Sensors	

Instrument Sheet

Manufacturer: ______________________

Instrument Name/Model No.: ______________________

General Information	
Hazard Type	
Instrument Life Expectancy	
Instrument Type	
Manufacturer	
Sensor Types	
Display	
Battery Type	
Dimensions-Stripped	
Dimensions with Accessories	
Low Battery Indication	
Weight (with/without Accessories)	
Warranty	
Approval Standard(s)	

Safety	
Hazardous Area Rating & Classification	
Ingress Protection Rating	

FIGURE 5.1
Standardized equipment specification sheet. ***Source:*** AIHA (2022)

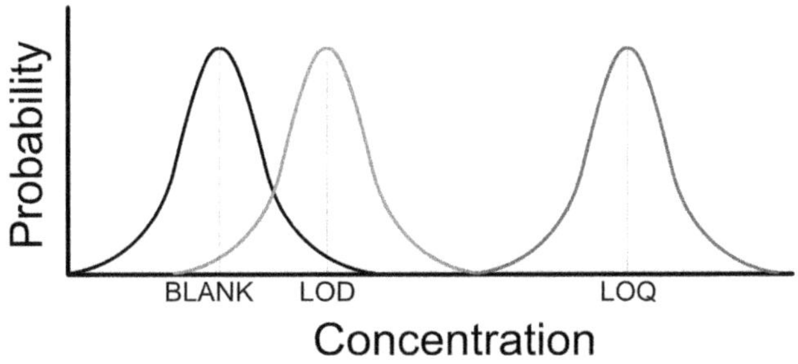

FIGURE 5.2
Limit of quantification and limit of detection. ***Source***: Image courtesy of Stephen Lee (2026)

detection limit, and quantification limit level samples using three standard deviations (3s) and ten standard deviations (10s) approach (Uhrovčík, 2014) (image used with permission).

Accuracy is the closeness of agreement between a test result and the true value (ISO, 2023).

Precision is the closeness of agreement between independent test results; it does not relate to the true value or the specified value (ISO, 2023).

Sensitivity, as it relates to analytical testing for exposure monitoring, is the probability of a sample having a positive value and it being the true value of what was detected in the field (McKay et al., 2021).

Selectivity is the ability of the detector to detect the contaminant of interest while in the presence of other contaminants (Ballantine Jr., et al., 1997).

Operating Limits or Range relate to the ambient or environmental conditions in which the device will be operated. For industrial work environments, this usually includes various pressures, humidities, and temperatures which may impact the sensor output and longevity. The Operating Limits are typically displayed as a range of physical properties in which within this range there is more confidence that the device will work as intended. Outside of these ranges, sensitivity and accuracy may be negatively affected.

Humidity is important to test and monitor for sensors, as moisture can interfere with the sensor mechanisms. Overall humidity refers to the relative humidity of the ambient environment. *Relative humidity*, or %RH, is a measurement of water vapor in the air that is compared to the maximum amount of water vapor that can exist in the air at its current temperature (NOAA / JPL, 2016).

Temperature for the purpose of our discussion refers to air or ambient temperature, which simply can be defined as the measure (in Celsius or Fahrenheit) of the average kinetic energy of moving air molecules, indicating how hot or cold the air is at a specific location. It represents the intensity of molecular motion, where higher temperatures mean faster-moving molecules and lower temperatures mean slower-moving molecules (Wikipedia, 2025). Like relative humidity, the ambient or industrial environment temperature will help in the selection of which sensors could function reliably in that environment, as indicated by the testing specifications. Especially if the sensor detects gases or vapors, operating temperature is critical to the accuracy of the sensor.

Interference or cross-sensitivity is particularly important to understand and look for when evaluating any sensor technology, especially when worn by workers who are required to respond to the alarms. Interference or cross-sensitivity is the tendency of a sensor to detect the presence of a different agent other than the target agent, leading to false readings. In gas detection, this is often seen with a hydrogen sulfide sensor picking up hydrogen gas in the area, or a carbon monoxide sensor picking up hydrogen gas (ISC, 2019).

Variance is the spread of data points from the mean. The higher the variance, the less predictability of the result; lower variance is an indication of good repeatability, precision, and reliability. Often, higher variance is seen when the sensor is being used in environments outside of the operating parameters. When evaluating sensor technology, selecting a device with low variance is preferred.

Case Study: Sweat Sensing Wearable (Biowearable)

All industrial sensors are developed in a laboratory setting. Today, there are many renowned universities with years of research and funding tied to sensor development. Professor John A. Rogers developed one such lab at Northwestern University (Rogers, 2023), expanding a research project to a wearable electronic device that analyzes sweat. The device started on a benchtop and underwent various forms of scrutiny before being accepted by the scientific community. Additionally, this sensor broke barriers in the sensor technology field, given that a sensor wasn't being used as a surrogate to exposure; rather, it was being used to measure the biological impact of exposure.

As biowearable sensors advance, the measurement of the "gold standard" must also advance, or rather evolve. Most lab bench measurements evaluate the physical nature of the sensor. As sensor technology shifts away from surrogate exposure measurement toward assessing physiological changes, the lab bench analysis will also need to consider how well the sensor reflects the physiological response(s) in a controlled laboratory or field environment using traditional measurement testing (e.g., EKG for heart rate). Biowearables introduce an opportunity for research, development, and validation standards that will be needed for biosensor acceptance within the scientific community and underscores the significance of technological advancements in health monitoring applications.

Epicore Biosystems, Inc (a biowearable sensor company) well summarizes an example of benchtop testing in its evaluation of biowearable technology

(Spinelli, 2025). Specifically, the benchtop calibration of the electrochemical sodium sensor, volume and flow sensor, and temperature sensors evidences the study authors' comparison of their sensor outputs to the known and accepted practices for measuring sodium, volume, flow, and temperature. An R^2 of 1.0 or near 1.0 indicates fit or the sensor's ability to read the true value in a controlled, laboratory environment. A good fit in a controlled lab environment is important, as adding inherent variability from a field environment can create unpredictability. When evaluating technology, both the benchtop and field testing should be considered.

6

Evaluating Technology: In the Field

Introduction

The laboratory is an easily controlled environment; the field, hardly. In the last chapter, we discussed a range of conditions, such as humidity and temperature, that a sensor or equipment may be exposed to, revealing how it would perform in those "operational conditions."

Variability

Variability in a device's accuracy or precision in an ambient field environment may increase, decrease, or parts of the device may not function at all. When evaluating technology to be used in the field, it's important to understand the conditions in which the sensor and housing were tested versus where they will be operated. If the sensor was tested and continued to work with acceptable precision and accuracy between 40 and 90°F, and the environment in which the device will be used is either below or above the tested range, the device's performance may not be as reliable. This doesn't necessarily mean the device is not usable; you simply need to understand what is affected and adjust for these effects when interpreting results. Some commercially available sensors and equipment may be tested under likely conditions where it performs slightly outside of the tested parameters while maintaining accuracy. Again, you would need to understand what is impacted and adjust the results accordingly. Even if the operational environment is within the lab bench tested parameters, variability is introduced from the quality of the material to the shipping and activating the device for the first time.

DOI: 10.1201/9781003564782-6

Value of Field Testing

Field testing can be thought of as "real-world" validation of a technology's performance. The reason many still perform quantitative analysis on a lab bench is that it's difficult to "shrink" the housing and design sensors that can withstand a wide range of ambient conditions. We see this in the initial design of industrial hygiene sampling, which is still considered the "gold standard" of occupational hygiene exposure monitoring. Field testing allows side-by-side trials of both the well-accepted analytical monitoring and the wearable sensor technology, revealing whether the potentially less expensive and real-time monitor provides repeatable information (where a correction factor could be applied) to enable quality decision-making for workplace health and safety programs.

Understanding Analytical Field Data for Industrial Hygiene Monitoring

When referring to analytical exposure monitoring, we are referring to a lab-provided media applied in the field, typically with a pump to collect concentrations of a vapor, gas, or particulate in the air, which settles onto a substrate in that media collection device through adsorption, absorption, or impact. Common substrates include charcoal or polytetrafluoroethylene (PTFE) that have been prepared using industry-certified and adopted standards in the laboratory. Placement of the device on the worker is considered, based on the movement of the contaminant of interest, as well as evaluating if the weather is conducive to sampling. Temperature, humidity, rain, and even sunlight can "spoil" the sampling media and produce inaccurate results.

Air with the agent of interest moves through the sampling media; the container is closed off and then transported to the analytical lab. In the lab, through standard operating procedures, the lab technician removes the substrate from the holding device and then removes the chemical agent(s) of interest from the substrate. The technician then follows standard procedures such as changing physical states through temperature or adding reagents to create a product that is then processed through the lab bench analytical equipment.

NIOSH (National Institute for Occupational Safety and Health) Methods are a great tool for prescribing the lab bench method for field sample analysis (refer to the bottom section of the method, where it describes, step-by-step, laboratory preparation before and after sampling). The lab method procedures and the equipment have undergone robust, well-established quality assurance and quality control criteria that span decades of repeatability. Within the industrial hygiene regulatory compliance community, these methods, OSHA methods, and lab accreditation are considered the "gold standard" of analysis and reporting. Combined, these factors are often considered to mean the reported value is as close to what was present during that field sampling campaign, and the differences are thought to be known by a specific margin.

Testing Viability of Direct-Read Field Data

The challenge for direct-read instrumentation is the ability to reliably maintain the same level of performance in accuracy and precision as the analytical sampling method performs in the field. The benefit of direct-read samplers and wearables is that the results do not require a laboratory and can be instantaneous.

Direct-read instrumentation and sensors are developed in labs, tested on the lab bench, and then (hopefully) trialed on workers in the field working in variable ambient conditions. An advantage of direct-read instruments is that many technologies utilize advanced algorithms to adjust the outputs based on various inputs. These adjustments can be thought of as "correction factors" that closely match the results that would be achieved through an analytical sample. When evaluating direct-read instruments for use by workers, you will want algorithm development to be founded on a clear understanding of associated inputs, such as location on worker, ambient environment (temperature and % relative humidity) or presence of other contaminants. It also drives the need within the industrial hygiene, health, and safety community to advocate for developing standards to help provide results around expected precision, accuracy, and tolerance within ambient conditions. When evaluating technology for adoption, it is helpful to review the results of field testing and evaluate those results from both the user experience and analytical perspectives.

Example Questions to Ask against the "Gold Standard"

- What is the variability of the results (accuracy and precision) when in the field environment?
- Which environments and ambient conditions were these devices tested in?

- What was the level of user fluency and comfort with technology?
- What resulting adjustments were made to the algorithm compared to what was programmed in the lab?
- Which algorithm inputs are directly measured vs. preprogrammed vs. obtained from an external source?
- What is the intended use case for these devices?
- How does the algorithm continue to learn based on continued use?

Example Questions to Ask Around User Impression(s)

- What is the overall user feedback? Will they use it?
- How well did the device connect over Bluetooth®?
- Were the users receptive to the alarms? Did they believe them? Were they considered a nuisance?
- Did the users align with the use case and see value in using the device?
- How easy was it to view and transfer the data?

Conclusion

Before adopting technology for the workplace, it is prudent to perform both lab and field testing. Many manufacturers will have data for both, and before reviewing their information, it's important to know the intended ambient conditions a device would be used in, the work environment (within a hazardous classified area), connectivity, the technology fluency of your workforce, and how often the workers would need to interact with the device. It's also important to ask yourself, "What would you evaluate when adopting wearable technology in the field?" "Who would you want to wear the device?" Most importantly, "How do you convince workers to wear the device and trust the information?" We'll cover user acceptance in the next chapter. One teaser I'll give you is that it starts well before you purchase any equipment.

7

Worker Acceptance

Introduction

A main differentiator of personalized health data is that the worker is given results and recommendations specific to their individual needs. Permissible Exposure Limits (PEL) are established by reviewing available studies, which typically evaluate exposure at the LD50 (median lethal dose) (Morris-Schaffer, 2021) of the studied population and then apply safety factors to account for unknowns (e.g., animal to human uncertainty, more sensitive populations, etc.). There is still the possibility that a worker may be more sensitive, may have comorbidities that were not accounted for, or may even experience amplified exposure due to another contaminant's presence, a sort of synergistic effect. For example, a recent study began to explore the toxicity of common workplace contaminants when the worker is also exposed to high temperatures (Leon, 2008).High temperatures increase workers' susceptibility to chemical exposure, a critical risk that standard PELs may ignore. Therefore, generalized exposure limits may not offer adequate protection. Providing workers with individualized data and custom mitigations that account for their specific environmental work condition(s) and as sensors evolve providing just-in-time feedback on their body's effect of the potential exposure is a far more effective strategy to ensure safety and encourage the adoption of new technologies.

The chance of developing a chronic illness may not always be the driving force behind workers' decision-making. Many work to provide for their family, support their lifestyle, survive, etc. There may also be a workplace culture of norms or an air of machismo, potentially dictating a higher risk tolerance for one's own health. Therefore, spending time to evaluate every risk factor, such as a potential comorbidity or contaminant that may lead to a synergistic effect, may not be at the forefront of mind or prioritized when thinking about a work environment. Many workers and employers want clear instructions on what to do to protect themselves and then get to work. Some workers have a strong working relationship with their health and safety department and have open conversations around potential exposures, tasks that are difficult

DOI: 10.1201/9781003564782-7

to perform, etc. While others may rely solely on a procedure, their experience, or their immediate crew to stay safe. Worker acceptance of any tool or technology is not without hurdles, particularly skepticism and the fear of data being used to impact one's employment.

Looking at the Technology Acceptance Model in Figure 7.1 (Marikyan, 2025), there are specific areas to focus on when deciding to bring any technology into the workplace, let alone a health wearable. Perceived usefulness necessitates a strong use case or business case for intervention. Worker adoption of technology is more likely when workers participate in building the use case, selecting the technology, piloting the technology, and creating the system to manage it. Without worker involvement and ownership, adopting technology at the work site may fail before it even enters the "door."

Field pilots often start with an "off-the-shelf" technology to test whether a hypothesized use case is valid. Take the sweat-sensing wearable case study as an example. The difference among workers in developing heat stress is based on the worker's individual activity and physiology, which may not be adequately identified or mitigated when using environmental ambient monitoring devices (i.e., "gold standard" Wet-Bulb Globe Thermometer—WBGT). To execute a field pilot, it's important to still use traditional and well-accepted practices for measurement while also applying the innovative technology. For detecting the potential for heat stress in a workplace, the same ACGIH and NIOSH calculations would be applied using data from the WBGT (temperature, humidity, etc.) and compare those same calculations using the individualized data from the biowearable. From there, it can be seen if there is a difference in data and possibly a population within the workforce that could benefit from further heat stress mitigation. When biowearable data results are shared with workers, this is when feedback around perceived usefulness is noted. Does the worker feel safer? Will they continue to wear it voluntarily after you leave? Are their field supervisors or managers asking for it?

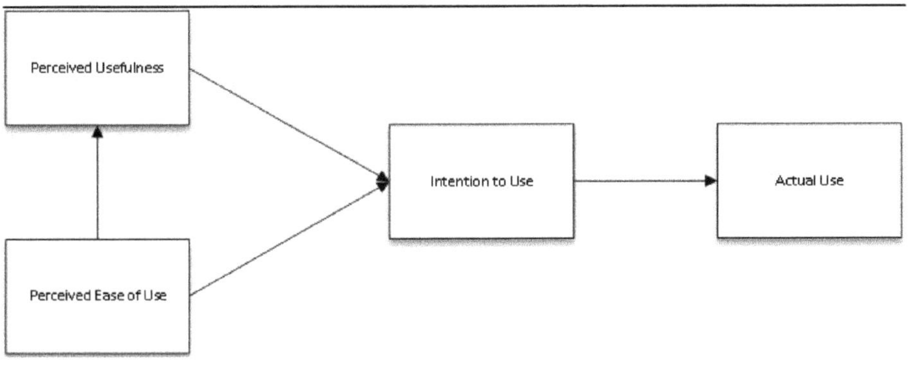

FIGURE 7.1
Technology acceptance model. ***Source***: Marikyan, 2025 (used with permission).

The *perceived ease of use* is identified by watching how users interact with the device. Does it cause a significant delay in their workday? Do they need to adjust their work clothing or personal protective equipment to wear it? Do they need to use equipment that they are not familiar with and requires a steeper learning curve? Another area to observe when conducting a field pilot is to understand where the biowearable would be most effective and where the workers would want to test it out *(intention of use)*. Each pilot and field trial will lead to more fine-tuning of the technology acceptance model, with the goal of creating a commercialized product that the workers would *actually use.*

Selling Technology

One purpose of performing pilot studies or field trials with workers and technology is to continually learn about the benefits and the pitfalls. It is also a wonderful opportunity to share learnings and experiences from the pilots and trials with a greater audience in hopes of volunteering for the next trial location. Pilot studies also allow consideration of other use cases that hadn't been thought of before. Gaining the "pull" instead of the "push" further informs industrial hygienists that pilot studies are needed and perceived as beneficial.

Measuring Worker Acceptance

How would you measure components of the previously discussed technology acceptance model to understand if the technology is a good fit? You could review compliance reports, such as duration of use, number of times the user interacted with the app, or trend the results to identify any improvements. You could also conduct pre- and post-questionnaires and interviews. Questionnaires are a great qualitative tool that gives the user an opportunity to think objectively about their experiences within the areas of interest. A questionnaire can also lead to more in-depth conversations through a follow-up interview with the user and project manager or design team. Workers' experiences, beliefs, likes, and dislikes are shared in these interviews, allowing for rich data that can be leveraged with leaders and other workers as you navigate adoption and deployment.

Leadership Culture

In addition to worker acceptance, it would be prudent to also touch on the importance of leadership endorsement and agreement. The workforce needs to feel supported by both their direct supervisor and organizational leadership in the sustained use of technology. When identifying use cases, it's critical to identify the issue that is being solved or enhanced and how it aligns with organizational goals for health and safety. Aligning with broader goals outside of the health, safety and environmental (HSE) function, such as digital adoption, cybersecurity, data science, and artificial intelligence, will further generate support for technology.

8

Digitizing Sensors

Introduction

In today's rapidly evolving technological landscape, the concept of digitizing exposure sampling is revolutionizing how individuals and organizations approach personal safety, environmental health, and occupational risk management. These devices digitize a technical method, developing an electronic sensor and form factor to continuously measure, record, and analyze a user's exposure to environmental hazards, offering unprecedented insight and protection. As industries and societies become more conscious of health risks that range from air pollution to radiation, these digitized sensors serve as the next frontier in proactive and personalized safety management.

Defining Exposure Monitor Wearables

Exposure monitor wearables are compact electronic devices, typically worn on the body, designed to detect and record exposure to an environmental agent of interest. These agents may include physical factors (such as noise, vibration, or radiation), chemical substances (like volatile organic compounds, gases, or particulates), or biological threats (such as pathogens or allergens). Traditionally, such monitoring relied on stationary or manual sampling, which often provided limited, time-specific data. Wearables, on the other hand, offer real-time, continuous monitoring tailored to an individual's unique exposure profile with the ability to pre-set alerts and alarms for thresholds.

The Process of Digitization

Digitizing exposure monitor wearables involves integrating digital technologies like microprocessors, sensors, wireless connectivity, cloud storage,

DOI: 10.1201/9781003564782-8

and data analytics into the devices' core functionality. This transformation enhances the devices' capabilities well beyond basic detection and logging. Key elements of digitization include the following:

Sensor Integration: Embedding highly sensitive digital sensors capable of detecting minute concentrations of hazardous substances or changes in environmental conditions.

Real-Time Data Transmission: Incorporating wireless connectivity (such as Bluetooth®, Wi-Fi, or cellular) to transmit data instantaneously to smartphones, computers, or centralized servers.

Automated Data Logging and Storage: Storing exposure data in digital formats, often on secure cloud platforms, to enable longitudinal analysis and easy access.

Data Analytics and Visualization: Applying machine learning and sophisticated data analytics to interpret large datasets, recognize exposure patterns, and provide actionable insights via intuitive user interfaces.

User Feedback and Alerts: Delivering real-time alerts and recommendations directly to the wearer or safety managers when hazardous conditions are detected, enabling swift intervention.

Key Applications of Digitized Exposure Monitor Wearables

The digitization of exposure monitors has broad implications across multiple sectors. Some primary areas of application include the following:

Occupational Health and Safety: In industries like construction, mining, chemicals, and healthcare, workers may be exposed to hazardous substances or conditions. Digitized wearables enable continuous exposure assessment, prompt notification of dangerous levels, and data-driven decisions to enhance workplace safety and regulatory compliance.

Environmental Monitoring: For individuals living in areas prone to air pollution, wildfire smoke, or other environmental hazards, wearable exposure monitors provide real-time data about air quality and personal exposure, helping users take appropriate precautions.

Medical and Public Health: Patients with chronic respiratory diseases, allergies, or immune vulnerabilities can use wearable monitors to avoid triggers and share exposure data with healthcare professionals, leading to personalized treatment strategies and better health outcomes.

First Responders and Military: Firefighters, soldiers, and emergency personnel often operate in hazardous environments. Exposure monitor wearables offer immediate feedback on toxic gases, smoke, or radiation, improving operational safety and mission effectiveness.

Benefits of Digitizing Exposure Monitor Wearables

The shift from analog or semi-digital exposure monitoring to fully digitized wearables brings a multitude of benefits:

Accuracy and Sensitivity: Digital sensors are often more precise and can detect lower concentrations of hazardous agents, minimize false negatives, and improve early warning capabilities.

Personalization: Data is tailored to an individual's location, activity, and physiological context, providing more meaningful exposure assessments compared to area-based monitoring.

Continuous Monitoring: Unlike spot checks or batch sampling, wearables provide uninterrupted tracking, capturing data during movement, rest, and various microenvironments.

Data Accessibility and Portability: Digital records are easily accessed, shared, and analyzed from anywhere, enabling remote consultations, regulatory reporting, or integration into larger occupational health systems.

Preventive Action: Real-time alerts empower users to take immediate protective actions, such as evacuating an area, donning respiratory protection, or seeking medical help.

Trend Analysis and Research: Accumulated datasets facilitate epidemiological studies, exposure modeling, and risk assessment at both the individual and population levels.

Technological Components

Several technological advancements underpin digitized exposure monitor wearables:

Miniaturized Sensors: Advances in MEMS (Micro-Electro-Mechanical Systems) technology allow integrating multiple sensors into wearable formats without sacrificing accuracy.

Low-Power Microprocessors: Efficient computing hardware ensures prolonged battery life and complex data processing on lightweight devices.

Wireless Communication Modules: Reliable data transfer using Bluetooth® Low Energy (BLE), Wi-Fi, or cellular networks facilitates seamless connectivity.

Cloud Computing: Secure storage and advanced analytics in cloud environments reduce the burden on local devices and enable scalable solutions for large workforces or communities.

Mobile Applications: Companion apps offer user-friendly interfaces, personalized dashboards, and easy access to historical data, alerts, and recommendations.

Challenges and Considerations

While digitizing exposure monitor wearables holds great promise, several challenges persist:

Data Privacy and Security: Protecting sensitive exposure and location data from unauthorized access is key. Robust encryption, user consent, and awareness of privacy regulations are important considerations.

Calibration and Validation: Ensuring sensors remain accurate over time, especially in diverse and demanding environments, requires rigorous calibration procedures and quality assurance protocols.

User Adoption and Wearability: Wearables should be comfortable, unobtrusive, and compatible with users' daily routines or work attire to encourage consistent use.

Battery Life: Prolonged operation demands energy-efficient hardware and intelligent power management to minimize downtime.

Interoperability: Devices and platforms should adhere to open standards for seamless integration with other health and safety systems.

To illustrate these concepts in practice, consider how you would digitize a colorimetric sample. The results for a colorimetric sample are a change in color. It could include a deepening of the hue or a different color altogether. Ways to digitize would be changing those color thresholds into an electronic sensor. In previous chapters, we learned about the many sensors that are on the market and available. The challenge would be to find a sensor that could detect the agent or physiological media (sweat, breath, blood, urine, tears, etc.) and then quantify the concentration and have it as a digital output. Once the hurdle of translating a physical substance into an electronic response is figured out, the next step is to translate and quantify that response into a digital reading.

Keeping in mind that all that work was performed on a lab bench, how do you know that it's going to work in the device housing, let alone the field? Ways to evaluate it would be to:

- Compare the output to the industrial hygiene or medical "gold standard." For heat stress, this could be a WBGT or heart rate. For benzene, it could be a benzene colorimetric detection tube or a urine sample.
- Understand the ease of use and acceptance by the worker, industrial hygienist, and the field safety representative.
- Worker's ability to submit data through an application on a smartphone.
- Overall coordination of procurement, shipping and handling, training, data submittal, and data analysis.

Much can be learned in that first pilot, and it will likely further fuel excitement and desire to advocate and push for the tested technology. In thinking back on past pilots that I've been a part of, I've seen the most value in working with the same group of workers who test out each iteration of technology. From the proof of concept to the digitized device. By doing so, the workers are invested in development and help to shape how the wearable looks, feels, and functions. Do the values need to be translated into pictures? Do the colors make sense? Do the values that they get help them in advocating for their health? Another area to observe when digitizing technology is to understand the amount of intervention needed; high intervention by the user and/or a safety field representative would not lead to a good outcome in an industrial setting. The best design is one that focuses on ways to empower the worker to use the device, easily understand results, and collect those results without the pilot team's presence.

Digitizing a device and connectivity for a device are not the same thing. We described digitizing as translating a physical substance into an electronic signal, or whichever sensor technology is used. Connectivity is when a wearable device (the hardware data) connects to an application (software) on a computer or phone. Connecting the two can be done through a cord that transfers the data and downloads it onto the computing platform where the software translates the information into usable information (hopefully). However, when thinking about wearables and transferring information instantaneously, we are really describing wireless connectivity, which is usually referred to as Bluetooth®. Regardless of physical or wireless connectivity, the act of moving data from a physical device to software introduces challenges around data privacy (personalized data is now easier to obtain), workplace IT authorization and security, having the right connecting inputs/outputs, and whether an electronic device like a cell phone or laptop is even allowed in certain industrial areas. Like digitizing, connectivity also must be tested. When testing wireless connectivity, it may include evaluating if the signal can move reliably through clothing, if there is interference from other equipment, the distance required to transfer data, and this is just to transfer data from the hardware device to the software; we haven't even considered transferring the data to the internet.

The Future of Digitized Exposure Monitor Wearables

Looking ahead, the trajectory of digitized exposure monitor wearables is one of increasing sophistication and integration. The rapid advancement of artificial intelligence and big data analytics will enable predictive modeling, such as anticipating potential hazards before they occur, and more refined personalization. As smart cities and connected workplaces evolve, these wearables will become key nodes within the broader Internet of Things (IoT) ecosystem, interacting with building management systems, environmental fixed sensors, and emergency response platforms. Innovations such as flexible electronics, bio-integrated sensors, and energy-harvesting technologies will further improve adoption and use. The democratization of exposure monitoring, making it accessible and affordable for individuals in both developed and developing regions, will contribute to healthier, safer communities worldwide.

Conclusion

Digitizing exposure monitor wearables represents a transformative leap in personal and public safety, enabling continuous, accurate, and actionable insight into environmental health risks. By leveraging advanced digital technologies, these devices empower individuals, organizations, and societies to manage exposure proactively, improve health outcomes, and foster a culture of safety. As the field continues to advance, collaboration among technologists, health professionals, regulators, and end-users will be essential to realize the full potential of this groundbreaking approach.

9

Data Privacy and Cybersecurity

Introduction

In the West, and increasingly in other parts of the globe, the concept of privacy is rooted in an individual's freedom. According to the Law Institute, "privacy is about control over identity, our thoughts and choices" (The Law, 2023). Over the centuries, governments in the West have concluded that privacy is a human right and integral to preserving human dignity.

Privacy Law

Laws such as the General Data Protection Regulation (GDPR (European Commission, 2024)) in the EU and similar in the UK, the California Consumer Privacy Act (CCPA (CA DOJ, 2024)) in California, USA, Personal Information Protection Act (PIPEDA (Office of the Privacy Commissioner of Canada, 2025)) in Canada, Privacy Act & Australian Privacy Principles (APP (OAIC, 2024)) in Australia, and Japan's Personal Information Protection (PIP (Government of Japan, 2023)) (just to name a few) stipulate how data should be handled. These laws narrow the scope of privacy to that of data privacy and each party's responsibility. Implementers of a health system with the capability to sense/measure physical parameters from individuals should be familiar with key parameters and their associated responsibilities.

Personally Identifiable Information

Personally Identifiable Information is an important definition because it is the linchpin of many data privacy laws. Understanding what it is and how

DOI: 10.1201/9781003564782-9

to use or not use it when you structure your safety program gives you, the HSE manager, flexibility in achieving a safer workspace while maintaining privacy for your teammates.

Personally Identifiable Information (PII) is any data that distinguishes or associates the identity of an individual, either directly or indirectly. Given the spectrum of information constituting PII, examples drawn from laws and guidelines will be presented. Examples of direct PII include name, address, social security or any government identity number, telephone number, and email address. Indirect PII by itself may be difficult to identify an individual, but in combination with other data, it makes it possible to identify individuals. Examples of indirect PII include gender, race, birth date, and geographic indicators (US Department of Labor (DOL), n.d.). Both types are considered sensitive information. It is the responsibility of everyone to safeguard sensitive information, and it is incumbent on anyone who has access to sensitive information to safeguard that information.

What About Measurement Data?

Wearable sensors can provide measurement data specific to an individual, and in many cases, the sensors can share information. Shared measurement data that cannot be associated with an individual (either because there is no direct or indirect identifiable information) does not constitute PII. For example, as part of a health program, you issue pedometers to your team. If the pedometer data is not associated with any individual because names, email addresses, gender, race, or any identifying information are not linked, then the pedometer data is not considered PII. In other words, if the measurement data is not identifiable, the measurement data is not considered PII.

Privacy Protocols

There are other considerations beyond handling PII. Additional statutes may apply in your organization or jurisdiction. For example, in the United States, the Americans with Disabilities Act (EEOC, 2000) and in Australia, the Australian Privacy Act of 1988 apply to all employees—not just those with disabilities. Employers are prohibited from making health-related inquiries of employees unless they are "job-related and consistent with business necessity." These policies and similar ones from other regions give guidance as to when health-related inquiries and health-related data sharing are permitted.

Access Control

Workers may be concerned that a supervisor will access health data measured by a wearable that is not job-related and use that information as a reason for dismissal. Some companies, therefore, have created policies where only medical staff, not a worker's supervisor, may access individual worker health data from wearable sensors. You should understand your organization's policy and safeguards for access control of individual data. Other organizations may have stricter policies on sharing individual data.

Data Sensitivity

Increasing use of wearables for consumer sports applications provides workers with examples of the types of measurements wearable sensors can capture. Some of the measurements are considered health data. When creating a safety program with wearables, it is important to understand the type of data being captured. This serves three purposes: to understand if the data is considered sensitive health information; to assess whether sharing/viewing that information is necessary for the workers' safety or job function; and to understand the workers' perception of how information is being used. For example, heart rate monitoring has become very prevalent in consumer devices such as smartwatches, some of which have the capability to alarm when they detect "anomalous" rhythms. A worker who has a smartwatch with this capability may find it acceptable to utilize it for personal use, but may be strongly opposed to the employer mandating wearing such a device and seeing that information, especially if that data can be used to discriminate against the worker.

Workers may trust health and safety staff more than their supervisors because health and safety personnel do not directly determine the workers' employment status. One way of building trust is to communicate to workers what specific sensor readings are measured and what information is being shared with supervisors. Many concerns may be alleviated if data capture and sharing are limited to only what is necessary.

Use of Personal Information

I remember sitting in a large room that could fit 30 people, outfitted with long tables. There was a projector screen at the front of the room. We presented the Connected Hydration Sensor Technology to a group of Health, Safety, and Environmental (HSE) managers who expressed interest in the technology to help mitigate worker accidents and injuries in the hot months. We presented how the sweat monitoring technology behind Connected Hydration was originally developed for elite athletes, and having received numerous requests from industrial hygienists to use this sweat measurement technology for workers, we built a product specifically for outdoor and industrial workers.

One of the local HSE managers in the audience immediately raised his hand and said, "I understand elite athletes want to win and why they would want to wear this, but a lot of workers are concerned about 'Big Brother.'" He went on to explain that the workers were apprehensive about being monitored by technology. He went on to say they even balk at having the gas monitors (that are required) recording exposure.

What the local HSE manager astutely understood is workers' sentiment about privacy and data use outside of their control. Unlike the consumer sports wearables use case, where the buyer uses a wearable step counter and heart rate monitor to track a weekend workout, at work, there is a power dynamic between worker and employer. Will the data captured from a wearable be used by a supervisor against a worker? How will the data be used?

What privacy safeguards are in place? The assumption that a consumer wearable can just be deployed in the workplace is nuanced because the power structure and motivation in the workplace are different from those for a personal consumer device used solely by an individual. From an HSE perspective, the fact of the matter is that no matter how useful technology is in keeping a workforce safe, it is not useful unless the workforce participates. A safer workplace isn't achievable without trust from the workforce. This chapter will explain why data privacy matters, what steps may protect data privacy and cybersecurity, and how HSE managers might successfully deploy a wearable in their organizations.

Navigating Personal Information

When developing a safety program, consider whether you absolutely need PII to achieve your safety objective and whether it pertains to the worker's job function. If you are considering a wearable sensor that measures personal data, consider how necessary it is to access data specific to an individual. Can you achieve the same safety objective without PII? If you saw sensor data not associated with individuals, would that be acceptable to help achieve a safe workplace?

Deploying Wearables and Successfully Maintaining Worker Privacy

Several factors may be considered when determining how stakeholders use data from a wearable. What is the sensor measurement? Who "owns" that data? Who would benefit from viewing the sensor measurements? Who should have access? How can all stakeholders be appeased and collaborate for a safer workplace where privacy is a strong consideration? As an example, we will share how the privacy policy and PII is handled when the Connected Hydration Solution was trialed and how that became a blueprint for incorporating a physilogical wearable monitor into safety programs.

During the early trials of Connected Hydration, company we were trialing with performed a digital privacy review. The Connected Hydration devices measured sweat volume, electrolyte concentration, motion, temperature and, optionally, could log GPS location when the app was used (typically only during work hours while physically at work). Wearable predicates were reviewed. The most common was a wearable gas monitor for confined spaces, which is stipulated in governmental standards such as the Occupational Safety and Health Administration (OSHA) in the United States and WorkSafe Australia National Standard (Worksafe, 1994) in Australia. These direct-reading instruments measure exposure to gases, vapors, and fine particulates suspended in air. However, in those cases, the wearable gas monitor detects worker exposure to a gas or material specific to the worker's company. Legal reviews suggested that the company owns the gas and the gas monitor is measuring gas. In contrast, Connected Hydration measures a worker's motion, temperature, and sweat loss, telling workers how much to rehydrate. Given that Connected Hydration measures the worker's physiological data, legal reviews suggested that the workers own their own physiolgical data. Following through with this assessment, the Connected Hydration Solution is designed such that the workers own their own data and have the ability to delete that data at anytime. The Connected Hydration Solution's privacy policy states that by default only anonymized group statistics, such as number of alarms within a group of workers, are shared with supervisors and medical staff. Individual worker data is only seen by the individual.

Health and safety teams may be interested in using data to improve hydration and mitigate the risk of heat illness. To satisfy that request yet maintain worker privacy, Connected Hydration shows anonymized group statistics on the web portal. There is no individual identifying information, and the group statistics shown on the web portal would not constitute PII. The health and safety teams can observe group trends such as whether the group is hydrating more often or adequately, or if temperatures and work exertion are changing. They can improve training or address issues at a group level, all while maintaining individuals' data privacy. Protecting individual data helps mitigate the risk of violating employee work statutes because supervisors are unable to identify and discriminate based on an individual worker. In terms of privacy controls, employers do not have access to individual data and therefore are not at risk of discriminatory accusations from labor laws. Further, since PII is not provided, it does not fall on HSE or supervisors to manage it. In this example, a solution at the design level was devised that maintained individual worker privacy, protected PII, and reduced employer risk of violating labor statutes.

The example of deploying Connected Hydration showed that introducing a wearable that measures physiological signals required extra thought when developing policies and privacy controls. It also showed that there is a way to effectively incorporate wearables into health and safety plans without compromising PII.

How would you implement your wearable program?

Given your safety goals, the sentiment of your workers, and what you've learned about privacy laws and employment/labor guidelines, how would you implement your program? We offer a checklist of questions for your consideration.

- *What is your goal when deploying a wearable sensor?*
- *What is the mechanism that the sensor provides to attain safety? Is it an alarm to the worker? Is it an alarm to HSE and the facility where some action needs to be taken, such as group evacuation?*
- *What sensor measurements are logged? Are these sensor measurements considered PII?*
- *What are corporate policies and regional employment policies? Are there stipulations on asking health-related questions that do not pertain to the task?*
- *What is your access control to the data? Can all employees see data? If the data is medical in nature, e.g., blood pressure, is it only limited to medical personnel? How do you ensure access control?*
- *Is there a way to achieve your goals while minimizing access to sensitive data?*

As shown in the Connected Hydration example, the system is used to alarm, inform, and educate workers directly. HSE can only access anonymized group statistics and can provide programs and education for the group, rather than single out individuals.

Cybersecurity

Once you have decided how you will implement your data policy and understand what types of data you are handling and potentially processing, you can take stock of what protections should be in place. Your company may have a cybersecurity team. Often, the corporate IT group is responsible for cybersecurity. The goal is simple—keep data safe and secure—for the myriad of reasons listed previously.

Your cybersecurity team may want to understand the following areas:

Information Protection Safeguards

How is data managed? Is it Software as a Service, where the wearable sensor company manages all the data processing and houses the data? Or is the data processed and managed at your facility? Do you or your solution provider have information protection safeguards that comply with industry standards, such as International Standards Organization for Standardization ISO 27001/27002, NIST 800-53, or Service Organization Controls SOC2 Type 2? Your cybersecurity team may provide you with a questionnaire to understand how data is managed and the safeguards you or your solution provider have put into place. In addition, your cybersecurity team may require attestation or compliance to known standards demonstrated by a third-party audit.

If your wearable sensor data is part of a Software as a Solution (SaaS) service, the SaaS service provider may be required to obtain attestation of compliance with industry standards. For example, your company may require SaaS service providers to obtain an ISO 27001/27002 or SOC 2 Type 2 report. The report must be conducted by a third-party auditor, and the auditor must attest that the wearable company's solution complies with the best data-security practices and has adequate safeguards in place.

Security Compromises

1. *Reporting and Advisories*: Your cybersecurity team would want to understand how urgent security news is conveyed. What is the process for security updates? How long after a security compromise would news be delivered?
2. *Incident Response*: In the event of a breach in security, what is your and your data provider's incident response plan? What is the chain of command, and what is the pertinent contact information?
3. *Cooperation*: When investigating a security compromise, it is imperative that you and the solution provider have access to and make available records, logs, and information required to comply with applicable laws, industry standards, and/or company policies.
4. *Correction Obligation*: Typically, cybersecurity assessments will request corrective action to fix a security compromise, vulnerability, or breach.

Cybersecurity Assessment before Rollout

Fortunately, there are established processes to help ensure healthy cybersecurity practices. If your wearable sensor solution provider houses and processes all the data because it offers SaaS, then the responsibility primarily falls on them, and they would need to address your cybersecurity team's requirements. Your wearable sensor solution provider can answer your cybersecurity team's questions, and you can put the two groups in touch. In contrast, if you are directly managing the data and processing it, you would have to meet your cybersecurity team's requirements. Additionally, if your wearable sensor requires corporate network access, you will have additional questions and requirements to meet. One way is to have a SaaS solution where the solution provider manages cybersecurity and PII, if there is any.

To provide a concrete example, Connected Hydration is a software-as-a-service (SaaS) solution. Data is processed and housed by Epicore Biosystems. Epicore has successfully completed several years of SOC 2 Type 2 audits and supplies the attestation report compiled by a third-party, reputable, independent auditor. For HSE teams that use Connected Hydration, the process is straightforward; Epicore Biosystems provides the SOC 2 Type 2 report and makes representatives available for any cybersecurity questions. The questions may take place through a survey or be conducted live with your cybersecurity team directly.

Since data is processed by Epicore (company that developed the Connected Hydration Solution) and housed by Epicore, corporate cybersecurity teams would look to them to ensure compliance . Typical questions are then covered by general cybersecurity policy, such as password rotation and checking who internal to the HSE organization has access. In addition, a SaaS solution is often more streamlined since it does not require access into corporate IT infrastructure.

Like designing your safety program using less PII to increase worker buy-in and reduce liability, it is possible to reduce PII and reduce your team's cybersecurity overhead. As previously mentioned by example, since Epicore only shares anonymized group statistics with HSE (i.e., supervisors cannot pinpoint individual measurements), there may be less risk of a supervisor discriminating against a worker, but there may also be less PII for HSE to protect, reducing HSE's cybersecurity requirements. In this example, the burden of maintaining cybersecurity falls heavily on the SaaS provider, Epicore.

As the architect of your safety program, how you safeguard your data is important. Understanding data privacy and cybersecurity concerns and establishing your safety program to address those concerns upfront can make deployment much easier, keep everyone's privacy protected, and reduce necessary IT compliance.

Conclusion

Taking time to understand privacy, PII, and how to protect it would give you more freedom when developing and deploying a safety program with wearable sensors. NIST document 800-122 "Guide to Protecting the Confidentiality of Personally Identifiable Information" provides best practices in protecting privacy, but also identifies what information is private. Ask yourself questions about reducing the amount of PII needed while still maintaining an effective safety program. Reducing PII can help mitigate confidentiality and privacy concerns, providing more flexibility and benefits around data use and analysis. Cybersecurity requirements may be reduced if you don't have sensitive information. Workers are more likely to participate if you do not have their sensitive information. There are many ways to set up safety programs, and each work function and job site is different. We hope our stories and our experience are helpful when setting up the framework for your safety program using wearable sensors.

10

UI/UX

Introduction

Industrial hygienists and health and safety specialists involved in workplace design, inherently safer design, or human factors/human engineering likely have more experience in user interface (UI) and User Experience (UX) than is realized. If you've ever been consulted on the height of a valve or a sight glass, alarm sounds or colors, or facility signage and labeling, you are already ahead of the pack! In the case of wearables, UI refers to the specific digital screen or controls allowing the wearer to interact with the device. UX, in the case of wearables, refers to the user's overall experience with technology. When evaluating a technology, both the UI and the UX are extremely important for the technology's success. HSE field specialists cannot always be in the field when work is being performed. Therefore, it's important that the worker can operate and troubleshoot a wearable with minimal intervention from another party. If a wearable requires daily oversight, then it's not an effective wearable solution, and is instead a traditional industrial hygiene sampling technology. The return on investment in wearable technology comes with no longer sending technicians to perform frequent employee monitoring. Instead, workers are monitored from a distance, with data reported in real-time and available from wherever the internet may be accessed.

User Interface

When thinking of attributes of user interface, a good philosophy is to use practices that are part of mainstream technology. Remember the first time you had to learn how to "swipe" instead of pressing a hard button? The learning barrier to overcome that change didn't happen overnight and took time until it became second nature. Once a trend takes hold, it's hard to miss. Toddlers today know to swipe and not press buttons when grabbing a

DOI: 10.1201/9781003564782-10

smartphone. Therefore, you would want to adopt technology where the user doesn't need to teach themselves new ways of interacting with a screen. A good rule of thumb is to keep the user/device interaction to a minimum; the more you can automate, the better the user experience and the more successful the technology use (Ruddro & Mohana, 2023).

Design

Other user interface considerations are in design—color can be used to relay information quickly that a response is or is not needed. Using red, yellow, and green is often helpful to respectively indicate "stop," "caution," and "proceed." Color schemes are particularly helpful when a device's goal is to remove ambiguity and allow for quick decision-making (Nagare, 2024). Another consideration should be the shape and size of the device. Depending on how often the user would need to interact with the device, buttons should have a different texture than the housing, be large enough to press with a fingertip, and have a certain amount, but not too much, force to activate the button. Another concept to consider is enabling multiple ways of notification. Imagine an industrial setting where the environment has elevated noise levels, extremes in temperature or humidity, and light levels that could vary, with some areas being dim. The device could be under clothing and not accessible. An alarm that is audible may not be the best solution. Consider other physical sensing, like vibration, as a way to alert or alarm a user. There is a lot that goes into the user interface, from a safety and ergonomics perspective. The involvement of a product industrial designer or human factors engineer is a great indication that the developers maintained the user as the focus of the design and considered the safety elements and human response to design (Methlagl et al., 2023).

Comfort

Another critical area to explore during field pilots is how workers felt about device placement. Study populations can vary in skin sensitivity and what is considered an annoyance. Also, experiences of comfort can change as time goes on and with daily use of a device, especially if it's directly applied to the skin of a worker. To protect the reputation of any study, it's important to address any user concerns during a pilot immediately and relay that use case to the design team so they can explore other ways to maintain usability of the device.

Technology Fluency

Not everyone in a worker population is the same, and it takes trial and error to uncover which method or design works for most. Our study participants ranged in age, and that variability resulted in differing feedback related to user experience. We noticed the younger users could quickly interact with the device, some without training, while our older users required some initial hands-on support. For all age groups, after a few consecutive days of using wearable technology, they were pros! This was seen in how quickly users set up the device each day, as well as compliance, or rather adherence, to the device's recommendations, which we saw through active interaction with the application and device.

User Experience: Trusting the Data

There may be instances where a user does not believe or trust the data. As any good researcher, it's important not to dismiss the concern and to demonstrate curiosity and have a questioning approach to the algorithm that is providing results. These pilots are not only to test the user experience; they also gather data to continuously evolve the algorithm, so the algorithm and results could be incorrect. Trusting a device is incredibly important to its adoption (Methlagl et al., 2023). The goal of a biosensor or biowearable is to measure, monitor, and provide feedback that is accurate, so understanding if the device represents reality is an important measure of success. By keeping that open mind, the pilot creates an environment or culture of curiosity and learning both for the reseacher and the workers who are part of the pilot. Some may change the mindset they started with, as more data is generated and may become more aware and accepting of how well the results compare to their reality. That experiential data is priceless and will only help in establishing further trust in the data.

Gamification

There are many benefits to healthy competition and gamification in the workplace (Herrera et al., 2023). Once workers can start to see and believe their results, it's only natural for them to begin comparing themselves with each other. To help maintain an apples-to-apples comparison and to allow the data to be displayed, try to find a result that doesn't include sensitive health data and encourages the behavior desired to achieve the use case. Thinking about a sweat-sensing wearable, one metric to compare could be the workers' ability to maintain their baseline (staying out of a hydration or electrolyte deficit). I've deployed biowearables where a sense of camaraderie and excitement is built amongst the workers; realizing the impact and influence of a social network on the adoption and effective use of a wearable is something to consider when deploying new technology. Look for indications of gamification when participants break for lunch or sync their phones; are they voluntarily sharing their results with each other? With gamification, there is a balance between privacy and transparency of information, as well as ensuring the information is weighted for the population and promoting the right behaviors (Zhang et al., 2021). Overall, getting excited about taking care of themselves was a "win" in my book, and exploring gamification as part of the user experience may drive sustained use and adherence to the safety and health recommendations (Windasari & Lin, 2021).

Conclusion

UI/UX is paramount to user acceptance of technology, especially for workplace use. The aim when adapting a solution to a workforce is to remove as much friction from the system as feasible, applying from device delivery and ordering to training and feedback. One of the main goals of adopting wearable technology is to reduce technical personnel's reliance on sampling, allowing UI/UX to become the focus for successful adoption and ongoing use.

11

Network of Sensors

Introduction

In action movies, the protagonist at times has a sensor measuring his physiology. When trapped by the diabolical antagonist, the hero's biostats flash on control room monitors for his team to view. Data is easily accessible, and we can see how our hero is in trouble.

The rise of wireless networking, along with the miniaturization of electronics, has made these scenarios possible, not just in fiction but in real life. Except now, the protagonist is your industrial worker. What the movies never show is the amount of staff and budget needed to build up or support this network infrastructure. Nor do movies discuss the responsibility and liability of collecting this information.

We will step through the process of understanding how amalgamating information from many sensors can potentially help you save time in your job. We will first review a brief history of wireless networking and the "Internet of Things" (IoT). Then, we'll consider a few examples where networked sensors could save time for tasks, such as reporting.

History

In the early 2000s, wireless networking became widely available for home and office use. Industrial wireless protocols that were germinating for years, such as Zigbee, also saw some adoption. Useful communication between computers that showed value in the 1990s, such as email and searching for answers on a website, could analogously now happen at the instrument level. However, all was not so simple. Many of the existing instruments, such as hydrogen sulfide (H_2S) gas monitors, were designed for stand-alone use. Some of them gained connectivity options. While they were able to send measurement data or alarms, initially, there wasn't a straightforward way to analyze or even access that data.

DOI: 10.1201/9781003564782-11

The industry was confronted with a chicken-and-egg problem. Existing sensors weren't networked, and there was little incentive to network them for communication and data compilation. Innovative solutions that provide a networked solution need to establish their value. It was another 15 years until "the Internet of Things," an environment where devices such as sensors could exchange data with other devices over a communication network (Gillis, 2025) (International Telecommunication Union, 2012), reached cultural relevance. One notable characteristic of "the Internet of Things" is that devices automatically process, analyze, and perform functions without human intervention. An example is a networked irrigation system that monitors soil moisture levels, weather forecasts, and crop requirements to automatically adjust watering schedules (Gillis, 2025). In industrial settings, one common use of IoT is for asset management, where sensors are placed on equipment to track its location and movement. Further enhancements made possible by software that can aggregate data and present only what is needed easily and when needed, facilitated managers' ability to solve mundane tasks, like tracking equipment. With opportunities abounding for new capabilities in safety, it's helpful to consider the constraints of pertinent technology.

New Possibilities

Automatic sensor communication, aggregation, and analysis of sensor data open new possibilities. In the industrial environment, advancing sensor technology could help ensure compliance, better control operations, improve worker safety, increase machinery uptime, and accelerate response time with real-time data collection, data processing, and alerting.

As an example, OSHA dictates substances' Occupational Exposure Limits (OEL), such as hazardous gas levels or particle counts. In the past, instruments have merely provided a direct read of the measurement. Today, many of those instruments offer network capability and can perform additional functions, like aggregating data, logging it, and providing related records to industrial hygienists. This information, in turn, helps IHs understand trends and more efficiently maintain compliance records.

In some industries, monitoring exposure is mandated by law. For example, OSHA's Ionizing Radiation standards often require employers to monitor radiation exposure by measuring both in the environment and on an individual. Personal dosimeters are issued, and the dosage is tracked. Traditionally, a periodic dosimeter badge (UT Health Houston, 2025) is issued to a worker and collected after a pre-determined period. That badge is measured with an external instrument in a lab, and the radiation dose is logged and compiled. The process is manual in every sense of the word. Real-time radiation dosimeter patches (Mirion Technologies, INC, 2025), in contrast, can measure the radiation right on the badge. It can determine compliance (i.e., that the worker is wearing the badge). It can also detect radiation as it occurs and pinpoints the exact moment it is at a maximum level. Lastly, given that the measurement (and often the displayed reading) is shown on the badge, there is no need to submit the badge for further processing. If the real-time dosimeter is networked, the data can automatically be logged, saving additional effort by periodically collecting badges for further processing and reducing paperwork for subsequent logging (Siegel et al., 2019). The network of real-time dosimeters can automatically compile the workers' radiation doses, document compliance, and even flag issues such as when radiation exceeds mandated safe levels.

If you must manually read values from individual sensors to file a report monthly or quarterly, then you may be interested in having the network amalgamate and compile the relevant data for you. In addition, if you require your workers to meet compliance regulations by logging values, the sensors may automatically perform that function for you. Let's take the example of the networked, real-time dosimeter; the dosimeter logs which workers wore the device and when. The network can amalgamate the data detailing who is complying, whether anyone potentially exceeded the pertinent exposure threshold, and whether any devices require maintenance. All this information can be summarized into a report.

If there is an opportunity to save effort in logging and tracking, then networked sensors may be of interest. Let us learn how networked sensors may potentially be used to execute job duties.

What Kind of Data Do You Need?

Let us explore potential considerations when deciding what solutions may work best for your organization and supporting your IH or health and safety program. There are trade-offs. While gathering as much data as possible may seem useful at the outset, a plethora of information combined with increased monitoring capabilities may come at a cost: a more complex network, added responsibilities, and the potential for added liability. The following exercise steps through the thought process when figuring out what kind of data is needed. We will use the example of ionization radiation.:

What problem am I trying to solve?

Going back to the example of ionizing radiation—this activity is in part driven by regulatory standards for monitoring radiation exposure.

Can this be done with a stand-alone sensor? If so, how much labor does that entail?

Yes, exposure can be measured with a stand-alone sensor. This requires a process for issuing sensors, facilitating compliance to help ensure they are worn, collecting the sensors, sending those sensors to a lab, and filing a report. Would a networked real-time dosimeter save time? Yes. A networked, real-time dosimeter can automatically track usage (helpful for compliance) and summarize the data for reporting without requiring shipment to a lab.

If a networked sensor is worth considering, then what infrastructure is required? Do you already have that network at your facility?

Considering the ionization radiation example again, the real-time dosimeter requires a connection to a wireless network. A network requires ongoing maintenance. The next section will provide an overview of wireless networks. Some of the discussed networks may already be in use at your company. Understanding their strengths and weaknesses can help users determine how much data and what capabilities are preferred.

What is the most relevant data? What do I absolutely need? What are nice-to-haves?

These are critical questions to scope any project. Knowing what you absolutely need will help you settle on a solution.

Overview of Wireless Network Technologies

To better understand using a network of sensors, we will explain some of the infrastructure required and provide a background on the technology. In addition, we will review the pitfalls, what you should look for when utilizing a network of sensors, and the questions to ask when deploying a network of sensors.

Table 11.1 lists the most common wireless networks in industrial environments. The columns to focus on are "Common Use," which describes where they are commonly found or how they are typically used. The column titled "Max Range" lists the typical distance from a transmitter that a sensor can communicate. For example, for a mobile phone, we can view the first row that states "Cellular." On average, a mobile phone may communicate with a cellular tower from up to 40 km (about 24.85 mi) away. The maximum range values in this column are typical amounts, but they indicate how far the sensors can be placed from the transmitter. The column titled "Bitrate" shows how much information can be transmitted. To use an analogy, if data were water flowing through a pipe, the bit rate is the diameter of the pipe. A larger diameter lets more water flow through. Similarly, a higher bitrate means more data is transferred. We see that Wi-Fi has a bit rate of up to 1 Gbps (1 billion bits per second), whereas Bluetooth® has up to 3 Mbps (3 million bits per second).

What is not noted here is the power draw from the sensor to transmit at each corresponding rate and distance. It takes significantly more power to transmit over a cellular network over 40 km (about 24.85 mi) at 20 GBPS than it does utilizing Bluetooth® over 100 m (about the height of the Statue of

TABLE 11.1

Common Wireless Network Types, Technology Standards, Common Uses, and Characteristics

Wireless network type—colloquial term	Wireless network technology	Common use	Max range (meters, typical)	Bitrate	Note
Cellular	5G, LTE	Mobile smart devices for internet and communication, and unmanned vehicle connectivity	0.8–40 km	100 MBPS–20 GBPS	
Wi-Fi	802.11b, a, g, n, ac, ax	Home, office, and industrial computer and smart device networks	30–120	1 Mbps–1 Gbps	
Personal area network	Zigbee™ 802.15.4	Industrial automation, home automation	10–100	2–40 kbs	
Personal area network	Bluetooth®	Mobile phones, gaming, headsets, stereo audio streaming, smart homes, wearables, automotive, PCs, security, proximity, healthcare, sports and fitness, etc.	10–100	1–3 MBps	
Personal area network	Bluetooth® Low Energy	Mobile phones, gaming, smart homes, wearables, automotive, PCs, security, proximity, healthcare, sports and fitness, industrial, etc.	1–50	125 kbit/s, 500 kbit/s, 1 Mbit/s, 2 Mbit/s	
Wide area network	LoRaWAN	Smart tags for asset management, ID badges for smart buildings, and smart cities	12,000–15,000	980 bps–12.5 kpbs (limited to 11–242 bytes)	Lora is extremely low power and covers a large geographical area, but the data rate is slower, and the total amount of data is limited compared to the other wireless options.
Near field communication	Near field communication (NFC)	Digital wallet, embedded tags for passports, library books. Public transit cards	< 20 cm	424 kbit/sec	

Liberty) at 3 MBPS. Increased power means larger, heavier batteries or more frequent recharging.

Questions to Ask When Deploying a Network of Sensors

At this point, you have narrowed your scope to the type of data most relevant to you. You have an idea of what is necessary versus what is nice to have. Now, let us learn about the ramifications of the network. Table 11.1 and the following section will help you navigate and ask questions.

1. *What resources are required of my team?* Some of these networks communicate with a base station—such as wireless smoke alarms. Others require a personal network—like wireless headphones paired to a personal phone, creating a personal network. The type of network the sensor relies on dictates the expenses it will require. If the network requires a base station, do I need to purchase capital equipment to install that base station (like a wireless router in the home)? Does the network need to cover a minimum space to support a larger region? What is the network's transmit range? If there is a very wide transmit range, then many base stations aren't required. If the network also requires a personal area network, will it require a smartphone or tablet to connect to each sensor? Will workers need to provide a smartphone? What if workers do not want to use their personal phones? Do you already have a suitable infrastructure? In the case of the Wi-Fi smoke alarm, it communicates over Wi-Fi. Many industrial offices have Wi-Fi on premises. If you, in fact, have the infrastructure, are you permitted to access it?
2. *Are some of these network standards already commonly used in consumer applications?* For example, Bluetooth® is quite common among consumer devices. Therefore, most smartphones, tablets, and laptops are already Bluetooth® or Bluetooth® Low Energy capable.
3. *Liability and responsibility for monitoring data. Who is responsible?* Interestingly, networked sensors open possibilities for remote monitoring and alerting. It is possible for the sensors to alert a health and safety employee if the wearable sensor detects an anomalous signal. In some cases, such as for a lone worker laying pipe in remote geographical areas, it may be desirable to have a wearable that alerts for fall detection and when the wearer initiates a "call for help" action. For such emergency response applications, it is important to ask who is responsible for receiving alarms and for triaging. Is there a call center service that is provided with that networked sensor solution?

Will you and your staff be responsible for monitoring critical life-threatening events? What happens during hours when you are not on shift? Are you and your company liable if the communication network fails? Will the wearable sensor communicate wirelessly? Does your site have a networked infrastructure to ensure connectivity, and if not, is your team able to support the desired coverage for such critical situations? On the other end of the spectrum, there are wearable sensors that do not place the accountability of monitoring alarms on health and safety staff. Such sensors may be designed to transmit personal alerts to workers and log data. In the latter case, the responsibility falls more on the individual worker to act, and there is less reliance on the integrity of the wireless network. *Building or ensuring your network coverage will depend on the criticality of the sensing technology and how the associated responsibility is assigned.*

Putting It All Together

We continue discussing our case study of Epicore's Connected Hydration pilot to delineate relevant considerations when assessing needs and determining whether that wearable sensor's network communication suffices. Recall, Connected Hydration consists of a wearable that communicates to a smartphone/tablet via Bluetooth®. The phone/tablet communicates to the cloud via Wi-Fi or cellular.

Health and Safety teams use Connected Hydration to help mitigate heat illness incidents amongst their workers whom we refer to here as users. The wearable sensor measures sweat volume and electrolyte loss, in addition to temperature and motion. It also measures the amount of sweat lost and suggests the same amount of fluids and electrolytes for replenishment, which are displayed on the user's phone/tablet. The sensor's vibration alarm notifies users when they have reached critical levels of sweat loss without replenishing.

However, users are often not allowed to carry phones in their workplace, presenting an obstacle to the above-described sensor use. This highlights the importance of understanding your primary objective as well as your primary data. For Connected Hydration, the primary objective is to keep workers safe. They will receive vibration alarms even without their phones nearby. The system is designed so that workers can leave phones outside of work areas, such as in a break room, hydration station, locker room, or cabin. When they come back for a water break and check their phone, they can interact with the app to read their sweat loss, mark that they have rehydrated, etc.

Another issue arises in remote areas where Wi-Fi or cellular signals are spotty. Connected Hydration does not require a consistent phone connection. Rather, data is saved on the module, and that data can be transferred to the phone and relayed to the cloud when the phone reconnects to cellular or Wi-Fi service. However, this means the system may not truly function in real-time if the network fails. Again, this is where the scope of the project is important. Is it critical that every worker's sweat loss and alarm counts are visible on the cloud portal the second that information is detected? If so, ensuring network reliability is important. However, if the sensor is used more as a personal safety tool, then it may suffice that alarms and alerts are received in real-time by workers but group data is relayed within hours to the cloud for health and safety staff or team managers when the worker re-enters a cellular or Wi-Fi service. Ultimately, team leads and health and safety staff reviewed alarm counts, note trends, and used the aggregated group data to identified risk. They reviewed on a weekly to monthly cadence and followed up accordingly with groups of workers to help improve safe heat illness mitigation practices.

During early pilots, the topic of liability arose. If you can see critical alarms as they arise, to whom in your organization would that responsibility fall? Hospitals have a process where there are roles assigned to address critical situations. Nurses and medical technicians triage patients coming into the emergency room. Connected Hydration was designed as a behavior-changing tool to inform users about their sweat loss and rehydration practices. As such, the alarms go to the worker. Alarms are logged and transmitted to the cloud portal, but health and safety staff are not required to monitor for acute situations. Therefore, a spotty network (e.g., poor cellular) is tolerable. Given how the Connected Hydration wearable sensor is used by workers, the network communications required (Bluetooth Low Energy, Wifi/Cellular) was deemed sufficient for the needs of health and safety staff.

Conclusion

This chapter describes how amalgamating data from sensors can potentially save time, help with regulatory compliance, and provide new opportunities for improving safety. An overview of network technologies was provided. Finally, a case study explored various questions and conditions to determine if a network of sensors would be fit for purpose.

12

Software as a Service

Introduction

Software as a Service, or SaaS, is a cloud service and is a way of delivering applications over the internet instead of installing and maintaining software on local computers or in a localized network; you simply access the service online (Khalil & Winkler, 2023). This approach can make a significant difference for health and safety professionals. With SaaS, you can collect, store, and analyze health and safety data from anywhere, in real time. This means all your incident reports, compliance checks, and training records are centralized and easily accessible, reducing paperwork and minimizing errors.

Beyond just data storage, SaaS platforms often come with built-in analytics and customizable dashboards, so you can spot trends, generate reports, and make informed decisions faster. Updates and new features are rolled out automatically, ensuring your tools stay in line with current regulations and best practices, without an extra burden on your IT or safety team. SaaS can streamline health and safety programming, making it more proactive, efficient, and responsive to your organization's needs.

Ensuring a Smooth and Secure Adoption of Software as a Service

Define Your Requirements and Objectives

Before choosing or configuring a SaaS platform, clearly identify what you need the software to accomplish. For health and safety professionals, this might mean ensuring the platform can manage alerts from the wearable, compile user data, display external data, and generate analytics. Consider both immediate needs and future scalability. Engage stakeholders from IT, compliance, and the end user to outline all requirements and build an appropriate use case. Keep in mind that the desired data may require additional

DOI: 10.1201/9781003564782-12

hardware or energy from the wearable. The best practice is to keep the SaaS minimal and only utilize information critical to stewarding the program.

Evaluate Security and Data Privacy

Given that SaaS solutions store data remotely, security is paramount. Assess the vendor's security protocols, including data encryption (in transit and at rest), user authentication mechanisms, access controls, and regular security audits. For health and safety data, which may contain sensitive information, ensure the platform complies with relevant data privacy regulations (such as HIPAA, GDPR, or other regional or local standards).

Integration with Existing Systems

Consider how the SaaS will work with your current tools and databases. Does it offer application programming interfaces (APIs) or built-in connectors for seamless integration with HR, payroll, or compliance platforms? Smooth data flow ensures minimal disruption and avoids data silos.

User Access and Permissions

Define roles and permissions carefully to control who can view, edit, or manage specific types of data. For health and safety scenarios, you may need granular controls, such as restricting confidential reports to authorized personnel or allowing only certain users to approve compliance checklists.

Data Migration and Onboarding

Having the ability to manage datasets under one system of record is essential to data analytics and analysis. Understanding if a SaaS platform, which could also manage real-tiime wearable data, can upload or migrate existing data into that same platform would be a selling point for organization adoption. Ease of data migration would be based on evaluating data compatibility, cleanliness, and completeness. The onboarding process should include adequate training for users focused on signing in, data lookup, variability of data, detection of errors, and support channels to address initial questions or issues. Onboarding shouldn't be a heavy lift and a way to ease adoption is to find SaaS platforms with a UI/UX similar to common workplace applications in use.

Customization and Scalability

Not all organizations have identical workflows. Choose a SaaS that allows for customization, including custom fields, workflows, dashboards, or reports to suit your unique processes. Confirm the solution can scale as your user base or data volume grows, without degrading performance or usability. Since SaaS dashboards are accessible from the internet, data created in one part of the world may be viewed in another.

Reliability and Availability

Examine the vendor's Service Level Agreement (SLA) for uptime guarantees, support response times, bug monitoring and repair, and disaster recovery plans. For critical functions like health and safety management, high availability is essential to ensure uninterrupted access to vital data and systems.

Compliance and Regulatory Alignment

Ensure the SaaS solution can help you maintain compliance with industry regulations. This could include automatic audit trails, electronic signature support, and version history for documents. Periodic updates from the vendor are crucial to keeping up with shifting regulatory landscapes.

Cost and Contract Terms

Review the pricing structure: is it subscription-based, per-user, or usage-based? Understand what is included in the base price and what may incur extra fees (e.g., added storage, premium support, advanced modules). Scrutinize contract duration, renewal terms, and exit clauses, should you need to change vendors later.

Support and Vendor Reputation

Reliable customer support is vital, especially during the initial setup and rollout. Check the vendor's track record, available support channels, and responsiveness. Seek reviews, case studies, or references from similar organizations to gauge satisfaction and reliability. Will they provide initial user training or serve as a help desk for your staff?

Ongoing Updates and Feature Roadmap

A strong SaaS provider will offer regular updates, patches, and new features. Ask about upcoming improvements, how customer feedback is incorporated, and how you will be notified of changes. Do they have regular touchpoints and relationship building with clients to understand future use cases and needs? Provide you, as the contract owner, with a single client representative instead of a generic contact number? Leveraging a vendor that stays current and keeps a pulse on industry needs and technology advancements ensures your system remains secure and effective as needed to evolve.

Exit Strategies and Data Ownership

Plan for the possibility of switching vendors in the future. Clarify who owns your data, how it can be exported, and what happens upon contract termination. Understanding these terms in advance helps avoid "vendor lock-in" and ensures you retain control over critical information.

Conclusion

Setting up a SaaS solution for the first time requires a thoughtful, strategic approach. Considering and defining these critical factors, requirements, security, integration, user access, onboarding, scalability, reliability, compliance, costs, support, updates, and exit options will facilitate a successful deployment and user experience that enhances your organization's efficiency, safety outcomes, and ongoing adoption of future technology.

13

Artificial Intelligence and Machine Learning

Artificial Intelligence (AI)

Artificial Intelligence (AI) refers to the ability of machines or software to perform tasks that typically require human intelligence. For an industrial hygienist or health and safety (H&S) program manager, this means leveraging technology to analyze complex and large datasets, identify patterns, and even make recommendations that support safer workplaces. AI can process vast amounts of information significantly faster and more efficiently than a human, which is especially useful in environments where rapid and accurate decision-making is critical to worker safety.

Machine Learning (ML)

Machine Learning (ML) involves training algorithms to learn from data in various forms (structured, semi-structured, or unstructured) to improve analysis over time without being programmed for every scenario (Mohammed et al., 2016). In our context, ML is the engine that enables continuous improvement: it can adapt to new hazards, predict trends, and help us stay ahead of evolving safety challenges by recognizing patterns in incident reports, near misses, external data, or physiological signals from workers.

Usefulness of AI and ML for Physiological Sensor Detection, Interpretation, and Action

AI and ML are increasingly being integrated into the wearable physiological sensors platform; for example, to monitor workers' heart rates, fatigue levels, or exposure to hazardous conditions. These technologies interpret live

DOI: 10.1201/9781003564782-13

sensor data, flagging anomalies that might indicate health risks or unsafe conditions. For instance, if a worker's heart rate spikes abnormally during a physically demanding task, AI-powered systems can alert the user or even trigger a predefined response, such as stopping machinery or sending medical assistance.

Advantages of Using AI and ML for Wearables

The primary advantages for the use of AI and ML within a health and safety program include the following:

- *Proactive Risk Management*: An initiative-taking approach can detect early warning signs of potential incidents, allowing for intervention before harm occurs.
- *Data-Driven Insights*: This information offers the ability to analyze massive amounts of data that would otherwise be overwhelming, translating information into actionable strategies.
- *Continuous Improvement*: These systems learn and adapt, so our safety protocols can evolve based on real-world outcomes and emerging risks.
- *Resource Optimization*: Automation frees up valuable time for field safety representatives, industrial hygienists, and data managers, allowing us to focus on high-level strategy and worker engagement.

Disadvantages of Using AI and ML for Wearables

Despite their promise, AI and ML technologies have limitations:

- *Data Quality and Bias*: Poor or incomplete data can lead to inaccurate predictions or missed risks, and ML models can inadvertently reinforce existing biases that exist within the organization, industry, and government.
- *Cost and Complexity*: Implementing AI and ML solutions may require significant investment in both technology and ongoing training. It may require an initial time investment to baseline a wearable that uses AI and ML to enhance its algorithms. It may also require that wearables be assigned to individuals instead of a pooled use case, requiring the purchase of more equipment. Life cycle cost from the

natural resources required to run the AI or ML logic may soon be incorporated into the overhead of deployed technologies leading to an increase per unit cost.

- *Transparency*: Some ML models operate as "black boxes," making it difficult to understand how decisions are made, which may be a potential challenge to using the data for compliance, worker trust, and ongoing validation.
- *Change Management*: Integrating modern technologies can encounter resistance from teams unfamiliar or uncomfortable with automation.

Accepting AI and ML as a Viable Solution

To adopt these solutions confidently, industrial hygienists and other health and safety professionals must better understand how AI and ML work, including their limitations and ethical considerations. This involves ongoing education about data privacy, model validation, and how to interpret and use the discerning logic of AI outputs. Building trust with workers is also crucial, ensuring transparency about data use and decision-making. Collaboration with IT and data specialists will be essential to align these tools with organizational values, behaviors, and compliance obligations.

Training Machine Learning

What is the role of the H&S specialist in training ML for their use case? How much time should be allotted? Are time and effort worth it? How does it advance worker health and safety?

As industrial hygienists, our role in training ML models is pivotal. We provide the expertise needed to define relevant data, set thresholds, and validate model recommendations within the realities of our worksites. The time commitment can vary widely from weeks to years, depending on the infancy of the technology, complexity of the environment, and the availability of variable users and quality data. While it does require upfront effort, investing this time pays dividends in the form of more robust, responsive, and effective safety programs—you must go slow to go fast. By harnessing AI and ML, our investment does not become antiquated, and we can better safeguard our teams, reduce incidents, and foster a culture of innovation and continuous improvement in workplace health and safety.

14

The Evolution of Industry and the Role of Digital Wearables

Introduction

The journey of industrial evolution is based on significant milestones, each representing a leap in technological advancement, societal transformation, and what we would call today, disruption. For purposes of this discussion, we will frame technological disruption from the perspective of Carla Millar, Martin Lockett, and Ted Ladd:

> "Disruption in the context of technology and innovation is defined as "change that makes previous products, services and/ or processes ineffective." The implication is therefore one of discontinuity – previous technologies and/ or ways of working are no longer viable."
>
> **(Millar et al., 2017)**

From the dawn of Industry 1.0 to the emerging Industry 5.0, the integration of digital wearables has played a crucial role in shaping the future of work and human interaction with technology. At the outset of writing this chapter, our digital world has evolved to the point that Industry 6.0 is now being explored. What do all these mean, and how can the industrial hygiene profession take advantage of and/or shape the revolutions coming our way?

Industry 1.0 through 5.0—Through the Ages

Industry 1.0: The Age of Mechanization

Industry 1.0, also known as the First Industrial Revolution, began in the late 18th century. It was characterized by the mechanization of production

DOI: 10.1201/9781003564782-14

processes through the invention of steam engines and machinery. This meant shifting away from manual labor to assisted labor with machines run by steam (Gander, 2023). This era marked the transition from agrarian economies to industrialized societies, leading to mass production and urbanization.

Translation to Wearables: During this period, wearables were rudimentary and primarily focused on safety. For instance, miners wore helmets with lamps to navigate dark tunnels, which were an early form of wearable technology aimed at improving safety.

Industry 2.0: The Age of Electrification

The Second Industrial Revolution, or Industry 2.0, emerged in the late 19th and early 20th centuries. This period saw the widespread adoption of electricity, which powered new inventions such as the telephone, electric light, and internal combustion engine (Gander, 2023). The introduction of assembly lines revolutionized manufacturing, enabling higher efficiency and production rates. Communication also improved significantly both in business operations and for the greater society, such as long-distance communication.

Translation to Wearables: In this era, the development of wearables—like electric hearing aids—improved the quality of life for individuals with hearing impairments (IEEE, 2020). Additionally, wristwatches became more common, allowing workers to keep track of time more efficiently.

Industry 3.0: The Age of Automation

Industry 3.0, the Third Industrial Revolution, began in the mid-20th century with the invention and application of computers and automation. This era was marked by the development of digital technology, programmable logic controllers (PLCs), and robotics. Automation transformed industries by reducing human labor and increasing precision in manufacturing processes. The proliferation of electronics revolutionized the manufacturing industry with the use of sensors and microcontrollers to monitor and control machinery (Gander, 2023).

Translation to Wearables: The introduction of digital wristwatches and early fitness trackers, such as pedometers, marked the beginning of more advanced wearable technology. These devices allowed individuals to monitor their physical activity and time with greater accuracy. Within industrial hygiene, this is where we begin to see the use of personal digital gas detection monitors worn by the worker. Sensors and batteries are miniaturized; skepticism is widespread around the use of these monitors to characterize worker exposure, and comparisons between analytical and digital monitors persist.

Industry 4.0: The Age of Digitalization

The Fourth Industrial Revolution, or Industry 4.0, from the late 20th century to today, is characterized by the integration of digital technologies into manufacturing and production. This includes the Internet of Things (IoT), artificial intelligence (AI), big data, and cloud computing. Industry 4.0 has enabled smart factories where machines communicate and make decisions autonomously, leading to increased efficiency and customization.

In industrial settings, smart glasses are used to provide workers with real-time data and instructions, enhancing productivity and safety while reducing travel costs. Smartwatches and fitness trackers have become ubiquitous, with almost one in three adults reporting wearing a smartwatch or fitness tracker (Dhingra LS, 2023); the market is flooded with this technology and has become a common tool for casual health measurement. These devices not only track physical activity but also monitor heart rate, sleep patterns, and other health metrics.

For the industrial hygiene profession, this may look like the connection of multiple environmental sensors (fixed plant sensors or portable area monitors) and/or personal monitors, that are equiped with global positioning system (GPS) technology, overlayed on a geographic information system (GIS) spatial map. This technology would allow the IH to view potential exposures of workers in the work environment at any time the sensors collected data. This data could be extrapolated to model potential noise exposure, vapor clouds, particulates or aerosol distribution over time using x, y and z coordinates while incorporating the wind, temperature, congestion of the space and even the sun at the time of collection! For noise, the map could show the loudest equipment that could be mitigated by sound insulation, identify a specific work activity that may require additional hearing protection or help establish hearing conservation zones during the post-safety start-up of the plant. For vapor clouds, particulates and aerosols, it can support predicting which buildings need to be hardened to serve as a shelter-in-place or even help identify the source of a release. Some have suggested that personal biological and genetic predispositions could be incorporated into a wearable that would alarm when the worker enters a geofenced environment (NSC, 2024) which have a higher risk potential to exacerbate pre-existing conditions.

Industry 4.0 is here today, and the digitalization of manual activities supports the desire to have just-in-time or real-time data available to intervene when it is the most protective and beneficial for the worker. It does take a considerable amount of investment up front to establish these programs. What may contribute to the delayed or hesitant adoption by workplaces are financial and knowledge resource limitations as well as awareness of available technologies (Masood, 2020). However, by increasing the technical fluency of industrial hygienists and health and safety specialists, dedicating financial resources while building a team of teams culture, adoption becomes achievable.

Industry 5.0: The Age of Human-Centric Innovation

Industry 5.0 is an emerging concept focusing on the collaboration between humans and machines. Unlike Industry 4.0, which emphasizes automation, Industry 5.0 aims to enhance human capabilities through technology. This includes the use of digital wearables, such as integrated smart glasses, fitness trackers, and augmented reality (AR) devices, to improve productivity, safety, and well-being.

Real-World Example: In Industry 5.0, wearables like AR glasses and exoskeletons are being developed to assist workers in physically demanding tasks. For example, AR glasses can overlay digital information onto the physical world, helping workers with complex assembly processes. Exoskeletons provide support and reduce strain on workers' bodies, enhancing their capabilities and reducing the risk of injury (Vallée, 2024). This is where the Connected Hydration solution comes into play. It is taking a physical digital device, connecting it to a web-enabled device via Bluetooth®, which then transmits the data to a cloud-based database and algorithm. The cloud-based algorithm then translates the information into usable data for the user via a phone software application (App) so that workers may respond to their near real-time results to obtain better health outcomes and endurance. Additionally, the aggregated data can be accessed by any web-enabled device through a Software as a Service solution.

The Role of Digital Wearables in Industrial Evolution

Digital wearables have become an integral part of the industrial landscape, particularly in Industry 4.0 and 5.0. These devices offer numerous benefits, including:

- *Enhanced Productivity*: Wearables like smartwatches and AR glasses provide real-time data and instructions, allowing workers to perform tasks more efficiently.
- *Improved Safety*: Wearables equipped with sensors can monitor environmental conditions and worker health, alerting them to potential hazards.
- *Better Communication*: Wearables facilitate seamless communication between workers and machines, enabling quick decision-making and problem-solving.
- *Health and Well-being*: Fitness trackers and health monitors help workers maintain their physical and mental health, reduce absenteeism, and improve overall job satisfaction.

The future trends for wearables in Industry 5.0 are poised to revolutionize the way humans interact with technology and enhance various aspects of industrial work. Here are some key trends to look out for:

Advanced Augmented Reality (AR) and Virtual Reality (VR) Wearables

AR and VR wearables are expected to become more sophisticated, providing immersive experiences that can enhance training, collaboration, and productivity. For example, AR glasses could overlay complex schematics onto physical objects, aiding in assembly and maintenance tasks. VR headsets could simulate real-world scenarios for training purposes, allowing a large population of workers to receive expert instruction and practice in a safe and controlled environment.

Integration with Artificial Intelligence (AI)

Wearables will increasingly integrate with AI to provide personalized and predictive insights. AI-powered wearables can analyze data from various sensors to predict potential health issues, optimize workflows, and provide real-time feedback. For instance, AI algorithms could analyze a worker's movements and suggest ergonomic improvements to reduce strain and prevent injuries.

Enhanced Health Monitoring

Future wearables will offer more advanced health monitoring capabilities, including continuous tracking of vital signs, stress levels, and mental health. These devices can provide early warnings of health issues, allowing for timely intervention and support. Wearables could also offer personalized wellness programs based on individual health data, promoting overall well-being.

IoT Connectivity

Wearables will become more interconnected with other IoT devices, creating a seamless network of smart devices (Mamdiwar et al., 2021). This connectivity will enable wearables to communicate with machinery, vehicles, and other equipment, providing real-time data and insights. For example, a wearable could alert a worker to a malfunctioning machine or provide instructions for troubleshooting.

Exoskeletons and Assistive Devices

Exoskeletons and other assistive wearables will become more prevalent, helping workers perform physically demanding tasks with ease. These devices can provide support and reduce fatigue, allowing workers to lift heavy objects, maintain proper posture, and perform repetitive tasks without strain. Exoskeletons will be particularly beneficial in industries such as construction, manufacturing, and logistics.

Enhanced Communication and Collaboration

Wearables will facilitate better communication and collaboration among workers, especially in remote or hazardous environments. Devices like smart glasses and wristbands can enable hands-free communication, allowing workers to stay connected and share information effortlessly. This will improve teamwork and coordination, leading to more efficient and effective operations.

Customization and Personalization

Future wearables will offer greater customization and personalization options, allowing workers to tailor their devices to their specific needs and preferences. This could include adjustable settings for comfort, personalized interfaces, and modular designs that can be adapted for different tasks. Customization will enhance user experience and ensure that wearables meet the unique requirements of each worker.

Sustainability and Eco-friendly Designs

As sustainability becomes a priority, wearables will be designed with eco-friendly materials and energy-efficient technologies. This includes the use of recyclable components, solar-powered devices, and energy-harvesting technologies. Sustainable wearables will reduce environmental impact and contribute to a greener industrial landscape.

While wearables hold great promise for enhancing productivity, safety, and well-being in Industry 5.0, several potential challenges will also need to be addressed:

Privacy and Data Security

One of the primary concerns with wearables is the privacy and security of the data they collect. Wearables often gather sensitive information, such as health metrics and location data, which can be vulnerable to cyberattacks and unauthorized access. Ensuring robust data encryption and secure data storage is crucial to protect user privacy.

Integration with Existing Systems

Integrating wearables with existing industrial systems and infrastructure can be challenging. Compatibility issues may arise, and significant investments in technology upgrades may be required. Ensuring seamless integration and interoperability with other IoT devices and industrial equipment is essential for maximizing the benefits of wearables.

User Acceptance and Adoption

The successful implementation of wearables in the workplace depends on user acceptance and adoption. Workers may be resistant to using modern technology due to concerns about privacy, comfort, or the perceived complexity of the devices. Providing adequate training and addressing user concerns is important to encourage widespread adoption.

Comfort and Ergonomics

Wearables must be designed with comfort and ergonomics in mind to ensure they can be worn for extended periods without causing discomfort or fatigue. Poorly designed wearables can lead to physical strain and reduced productivity. Continuous improvement in design and user feedback is necessary to create comfortable, user-friendly devices.

Battery Life and Power Management

Wearables rely on batteries for power, and limited battery life can be a significant challenge. Frequent recharging or battery replacement can be inconvenient and disrupt workflow. Advances in battery technology and power management are needed to extend the battery life of wearables and reduce downtime.

Cost and Accessibility

The cost of wearables can be a barrier to widespread adoption, especially for small and medium-sized enterprises. Ensuring wearables are affordable and accessible to a wide range of industries is important for their successful implementation. Economies of scale and technological advancements may help reduce costs over time.

Regulatory and Compliance Issues

Wearables must comply with various regulatory standards and guidelines, particularly those related to health and safety. Navigating the regulatory landscape can be complex and time-consuming. Ensuring wearables meet all relevant standards and obtaining necessary certifications (e.g., intrinsically safe) is essential to their deployment in industrial settings.

Data Overload and Management

Wearables generate substantial amounts of data, which can be overwhelming to manage and analyze. Effective data management and analysis tools are needed to extract meaningful insights from the data collected by wearables. Implementing efficient data processing and storage solutions is crucial for maximizing the value of wearable data.

Ethical Considerations

The use of wearables may raise ethical considerations, such as the potential for constant worker monitoring and surveillance. Balancing the benefits of wearables with respect to workers' privacy and autonomy is important. Establishing clear policies and guidelines for the ethical use of wearables is necessary to address these concerns.

Technological Reliability

Wearables must be reliable and durable to withstand the demands of industrial environments. Technical issues, such as device malfunctions or connectivity problems, can disrupt operations and reduce trust in technology. Ensuring the reliability and robustness of wearables is essential for their successful deployment.

Conclusion

The evolution from Industry 1.0 to Industry 5.0 represents a journey of technological advancement and human ingenuity. Digital wearables have played a pivotal role in this transformation, enhancing the way humans interact with machines and improving various aspects of industrial work. As we move forward, the integration of wearables will continue to shape the future of industries, making them more efficient, safe, and human-centric. The future of wearables in Industry 5.0 is bright, with advancements in AR/VR, AI integration, health monitoring, IoT connectivity, exoskeletons, communication, customization, and sustainability. These trends will enhance productivity, safety, and well-being, making industries more efficient and human-centric. As technology continues to evolve, wearables will play a pivotal role in shaping the future of work and transforming the industrial landscape.

15

Looking Ahead

Recent Articles and Discoveries

A quick internet search of recent peer-reviewed articles around artificial intelligence (AI) and sensor technology provides some insight into the direction of this field. There appears to be continued advancement in biomedical sensing, improving the precision and accuracy of sensors and combining technologies to help create a more holistic and complete view or interpretation of an environment. Below are some examples of this work:

- Catalbas, M. C., Arı, F., & Mutlu, M. (2025). A real-time fatigue detection framework based on non-intrusive physiological sensing for human-AI collaboration in socio-technical work environments. *Cognition, Technology & Work*, 1.
- Even, A., Minderhoud, R., & Van Hoof, C. (2025). Measurements of redox balance along the gut using a miniaturized ingestible sensor. *Nature Electronics, 1.*
- Coloretti, V., Russo, G., & Cortesi, M. (2025). Reliability, accuracy, and ecological validity of a wireless pressure sensor for swimming performance analysis. *Sports Engineering, 1.*
- Khan, I., Anh, J.-K., & Kwon, Y.-W. (2025). Lightweight deep transfer learning for earthquake detection in resource-constrained IoT devices. *Journal of Seismology, 1.*
- Tobeiha, N., Memarian, N., & Ostovari, F. (2025). Optical humidity sensor based on G/GO nanosheets. *Scientific Reports, 1.*
- Shao, J., & Deng, L. (2025). Fault diagnosis and fault-tolerant control of position sensors for switched reluctance motors. *Journal of Power Electronics, 1.*

Recent sensor research continues to highlight several key trends that are shaping the future of sensor technology. These trends are driven by

DOI: 10.1201/9781003564782-15

advancements in various fields and the increasing demand for real-time information across multiple sectors.

One of the most significant trends is the integration of *big data and artificial intelligence (AI)* with sensor technology (Andronie, 2023). This combination enhances sensor functionality and expands the application scope. For example, emerging image sensors can process vast amounts of visual data, enable real-time decision-making, and improve efficiency. In autonomous vehicles, AI processes lidar (light detection and ranging) data to create detailed 3D maps and understand the vehicle's environment, advancing autonomous driving capabilities and safety features (Anderson, 2016). When thinking of disruption, this is where you can see sensors replacing activities that currently rely on human workers. From a safety perspective, this disruption would likely be supported, as it could reduce ergonomic issues from sitting, reduce driver injuries from motor vehicle collisions, and help with traffic flow!

Another trend is the push toward *automation and electrification*, particularly in the automotive and manufacturing industries. Sensors like radar, lidar, and image sensors are key to developing advanced driver-assistance systems (ADAS) and autonomous vehicles. These sensors provide safety redundancy, assisting in protecting drivers, passengers, and other road users. In manufacturing, the integration of sensors in robotics is a foundational technology for Industry 4.0, which combines traditional manufacturing with advanced technologies (UTI, 2024).

The *Internet of Things (IoT)* and *Industry 4.0* are also driving the demand for sensors. Sensors are integrated into various devices for real-time monitoring, predictive maintenance, and operational efficiency. Sensor manufacturers are focusing on improving precision, miniaturization, power, and cost-efficiency to meet the needs of applications in robotics, medical devices, equipment efficiency, and environmental monitoring (Elkateb et al., 2024).

Additionally, there is a growing interest in *biosensors* and *MEMS (Micro-Electro-Mechanical Systems)*. These technologies enable new applications in healthcare diagnostics, environmental monitoring, and consumer electronics. Biosensors, for example, are being used for real-time health monitoring and diagnostics, while MEMS sensors are finding applications in various fields due to their small size and high precision (Welburn L, 2025).

Overall, the sensor industry is experiencing rapid growth and innovation, driven by advancements in AI, IoT, automation, surface chemistry, and miniaturization. "The sector [supporting Industry 4.0] holds 2 million+ patents, with over 506,000 applicants contributing to a yearly patent growth rate of 3.59%. The United States leads with over 639,000 patents, followed by China with over 355,000 patents" (StartUs Insights, 2025). The sensor industry, coupled with AI, is expanding capabilities where the "Internet of Things" is evolving to "Internet of Everything," where trends show "*1.19% annual growth*, reflecting adoption as organizations pursue incremental automation complementing human skills rather than replacing labor" (R., 2025).

These trends are expected to continue shaping the future of sensor technology, leading to new applications and improved performance across various sectors (He, 2024).

With these advances and aspirational goals being worked on, the key challenges that we have discussed in this book remain. The future and sustainability of these technologies will depend on innovators' and entrepreneurs' willingness to tackle these challenges head-on. Let us remind ourselves of some of the key challenges still facing sensor technology.

Power Efficiency

One of the major challenges in sensor technology is power efficiency. Sensors, especially portable and wearable ones, require a reliable power source to function continuously and collect and transmit data. Battery life and energy consumption remain significant inhibitors to the advancement and optionality of portable sensor detection (Khan, 2016). With a demand for lower carbon and environmental stewardship, ethical and sustainable battery power sourcing will continue to be a factor in the selection of technologies.

Accuracy and Precision

Achieving high or "high enough" accuracy and precision in sensor measurements is another challenge. While detection technologies have evolved, there are still limitations in current accepted practices and our understanding of physics. The level of acceptance among the scientific community will continue to be debated and compared against a "gold standard." Technical bodies will continue to test methods for validation and performance characteristics (NIOSH, 2024). This level of consensus will be important to the future advancement and investment of these technologies to consistently be used to measure the effectiveness of health and safety standards.

Scalability

The ability to scale sensor technology for widespread use is a challenge. Access to scalable resources and infrastructure, such as hard-wiring and electronics, is necessary for the effective deployment of sensors (Soga & Schooling, 2016). Use of cellular and satellite technology is how sensor technology may be best deployed for greater coverage and transmission of data.

Environmental Factors

Sensors must be able to operate reliably in various environmental conditions. Factors such as temperature, humidity, and electromagnetic interference can affect sensor performance and accuracy (Farnell, 2025). When evaluating environmental or ambient conditions in industrial workplaces, it includes not only the environmental conditions like temperature and humidity, but also includes the specific tasks performed by the worker(s). Are workers in an environment with dust, water spray, chemical vapors, etc. that might interfere with the sensor reading accuracy?

Integration with Wearables

Integrating sensors into wearable devices poses challenges related to size, comfort, and durability. Wearable sensors must be lightweight, flexible, and able to withstand daily wear and tear (Zhao Y. W., 2025).

Data Security and Privacy

With the increasing use of sensors in IoT and wearable devices, ensuring data security and privacy will remain and require ongoing vigilance. Sensors collect sensitive information, and protecting this data from unauthorized access is crucial (Radanliev, 2025).

Cost

The cost of developing and deploying advanced sensor technology can be high, requiring various rounds of investor funding. Reducing the upfront and ongoing cost while maintaining performance and reliability will be a continuous focus area for sensor manufacturers (Bridgelall & Tolliver, 2021).

Interoperability

Ensuring interoperability between different sensor systems and devices is essential for seamless integration and data sharing. Standardizing protocols and communication methods takes planning and understanding the platform needs in which the technology would be deployed. Simply selecting an ecosystem will dictate requirements, limitations, and expertise needed to develop the interface and storage capabilities (Ahire, 2025).

Maintenance and Calibration

Sensors require regular maintenance and calibration to ensure accurate measurements. Developing self-calibrating sensors or reducing the need for frequent maintenance should be considered in the design (Clark, 2018). Even deciding to have an item shipped back for calibration or battery replacement requires some level of coordination with the end user. When considering hundreds of users, some in remote locations, shipping technology back and forth can be a barrier in and of itself.

Real-Time Data Processing

Processing vast amounts of data collected by sensors in real-time is based on precision needs. Advanced data processing techniques and AI algorithms are needed to analyze and interpret sensor data efficiently. Understanding the minimum amount of data needed to provide results with desired accuracy and precision is dependent on the application. Collecting data for the sake of collecting data is not always the best philosophy when you have minimal storage space.

These challenges highlight the ongoing efforts and research needed to advance sensor technology and improve its applications in various fields. As industrial hygienists, it is important to remember that understanding the barriers and limitations of any technology and accounting for them in your analysis is the most effective way to use these tools to enhance your workplace programs. Practices like peer reviews and the scientific method will remain critical as we wade through the massive amounts of information that will be generated from the use of these tools. For instance, in the case of AI as a tool, practitioners will need to show discipline and equip themselves with new skills to discern between real and phantom results.

Ongoing Research and Future Directions

Advancements in Sensor Materials and Miniaturization

Research is focused on developing new sensing materials such as graphene and nanostructured compounds to enhance sensitivity, selectivity, and durability. Miniaturization continues to drive innovation, enabling sensors to be embedded in compact, unobtrusive devices.

Integration with Artificial Intelligence and Machine Learning

The fusion of sensor data with AI and machine learning algorithms enables predictive analytics, anomaly detection, and intelligent decision-making, opening doors to new applications and efficiencies.

The Internet of Things (IoT) and Connected Ecosystems

Sensors and wearables are increasingly part of interconnected networks as seen in smart homes, industrial plants, and healthcare systems, which enable seamless data sharing, remote monitoring, and automated responses.

Personalized and Precision Medicine

The ability to continuously monitor and analyze physiological data brings the vision of precision medicine closer to reality, with treatments tailored to individual characteristics and real-time health status.

Conclusion

The future of sensor technology is marked by remarkable innovation, driven by ongoing research in AI integration, miniaturization, and advanced materials. Recent discoveries highlight the potential for more precise, accurate, and holistic sensing across domains such as healthcare, manufacturing, and environmental monitoring. However, the field faces persistent challenges such as power efficiency, data security, cost, interoperability, and user adoption. These challenges must be addressed for sustainable progress. The successful deployment and impact of these technologies will depend on the continued efforts of researchers, practitioners, and industry leaders to confront these barriers. As the volume and complexity of sensor-generated data grows, disciplined analysis and rigorous scientific practices will remain essential to making informed decisions and maximizing the benefits of sensor technology in real-world applications. At the crux of developing accurate sensing technology and real-time wearables is the desire and drive to make our workplaces safer, our workers healthier and reduce the data management burden that our industrial hygienists and health and safety practitioners must navigate. The occupational health and safety landscape is ready for disruption and with the establishment of guardrails, user acceptance and advancements in technology we will get there and be part of shaping Industry 6.0 and beyond.

Quick Reference Guide for Managing a Wearable Program in the Workplace

1. Identify a specific problem or issue that needs to be addressed. Consider hazard mitigations that may already have resources and funding, and identify if there is an opportunity to optimize or strengthen the existing program.
2. Understand how wearable technology would be used (work scope, working environment, shift duration, etc.) and how it would be integrated into an existing management hierarchy, or create one for your organization.
3. Scout the commercial and R&D landscape for solutions. Focus on the technology—the lab-based results, and if available, field-based results, and select one or two that meet the requirements for your use case(s).
   a. Technology specifications (often found on a specification sheet):
      I. Testing environmental limits for operation (humidity, ambient temperature, etc.).
      II. Testing of use in hazardous classified areas (flammable atmospheres, low oxygen levels, etc.).
      III. Battery—duration of a single charge, replaceable or rechargeable, frequency and ease of charging; skill set or certifications required to replace batteries, cost of battery replacement.
      IV. Connectivity.
   b. Technology Validation, Precision, and Accuracy
      I. Variability of the sensor—precision
         i. With and without the housing.
         ii. Lab and field application.
         iii. With and without user intervention.
      II. Testing against a "gold standard"
         i. How is this hazard traditionally measured?
         ii. What is used to validate well-accepted lab-bench or direct-reading instrumentation?
      III. Understand the algorithm.
         i. Which variables are directly read from the device vs. which use AI, machine learning, or external data to "fill

in the gaps"? Will the algorithm adapt or evolve based on data?

4. Identify the pilot population and metrics of success.
   a. What specifically are you testing, and how will you measure if you met, exceeded, or fell below the objective?
   b. When deciding your study population and design, you will want to consider the following:
      I. Workforce fluency with technology.
      II. Work environment (e.g., requires intrinsically safe devices).
      III. Workplace culture around health and safety.
      IV. Languages.
      V. Cultural norms.
      VI. Workers' hazard identification (ideally, you will want to pilot with the workforce that the use case is based on).
      VII. Supervisor/Leadership endorsement and support.
      VIII. Access to the work location (when testing proof of concept, you may want to have easy access (i.e., minimal security and presence on land).) When testing the commercialized or close to commercialized product, test the harder access locations.
      IX. Required certifications of equipment to be in the work environment.
      X. Will they want the results? What will be the mitigations you suggest based on the results (if in a centralized role, be sure to partner with your local leadership and HSE organization).
5. Connectivity and worker access to technology.
   a. Determine required access requirements (username and password, single sign-on, etc.).
   b. Manage and provide user notification for electronic data privacy.
   c. Work with your IT team to understand any cybersecurity risks and, if possible, embed the mitigations into the design.
6. Gain permission from users.
   a. Any collection of personal data requires consent either through a paper form or electronically.
   b. Develop a consent document that describes the study, its purpose, and how their information will be used.
   c. Data Collection Plan: Anonymized data will be the easiest and fastest way to expedite the study. However, if that is not possible, then work with the appropriate health ethics committees, like an Institutional Review Board (IRB), or your employer's legal team,

as needed. Have a plan in place on how to identify and store data, including validating that digital data was transferred and uploaded after each shift.

d. Perform a Data Privacy analysis and establish mitigations to protect collected data. Include a Data Privacy statement in the consent document.

e. Determine whether individual data will be shared directly with the trial participant; if not, be sure to add that to the consent document.

7. Work with the vendor/supplier/technical team to identify the number of samples necessary for statistically significant data. Coordinate visit with site team; identify any security clearances, transportation restrictions, escort requirements, worker work schedules/shifts to navigate and account for in the pilot study or trial.

8. Conduct trial.

   a. Meet with supervisors and site management to explain the technology, use cases being explored, and gain any further insight into needs. Inform them of the amount of time needed to onboard personnel, check-ins, daily don/doff of equipment, as well as follow-up questionnaires and/or interviews. Allow time for questions and answers.

   b. Meet workers who will be involved in the pilot. Take the time to understand how their work meets the use case that's being explored.

   c. Conduct and/or provide training materials and hands-on support for devices and onboard personnel as quickly as possible. Keep onboarding to at most a 1-hour duration.

   d. Have scheduled check-ins during the first few days of wearing the technology to assess comfort, review data, and make sure it's working!

   e. Allow a duration of time when the worker is left alone with the device from donning/doffing to turning on the unit and evaluating their data. This will help identify any friction in the system and where further support or research is needed.

9. Evaluate data.

   a. Summarize themes and trends.

   b. Evaluate data against use case measures of success and/or hypothesis.

   c. Provide individual workers with their data, if applicable.

   d. Generate a report describing the use case, trial specifics, number of participants, locations, measures tested, results, next steps, and conclusion.

10. Provide report to decision makers (e.g., R&D funding team, Digital Technology management team, HSE management, Operations Management, etc.) Evaluate alternatives or implement technology.
   a. Determine if, based on results, the tested technology will work, and then deploy.
      I. How was worker acceptance?
      II. Was the data accurate? Did workers, management, and HSE field specialists trust the data?
      III. Did it meet or exceed the metrics being measured?
      IV. How did the workers feel after using technology?
   b. Evaluating other technologies.
      I. Trial with the same user group, or a different user group.
      II. Evaluate against the same metrics of success.
      III. Develop a report comparing the technologies and provide a recommendation to leadership.
   c. Selecting and purchasing a solution.
      I. Become familiar with your workplace procurement and purchasing policies.
      II. Work with the technology manufacturer to reach an agreement around procurement and implementation.
         i. Provide them with an estimated number of units to determine the best unit price. Are there discounts at a certain threshold? Duration of use?
         ii. Cloud services and SaaS log-in users.
         iii. Support/Help desk.
      III. Site Deployment and if it includes collaboration with IT and Cybersecurity Subject Matter Experts (SMEs), if applicable.
         i. Ongoing check-in with solution provider and users; adjust as needed for most effective use of technology.
         ii. Participate in technology road map conversations with manufacturer to help shape future functionality and enhancements of solution.

References

Ahire, A. (2025, August 18). *IoT device testing: Interoperability testing for IoT: Ensuring seamless protocol & device interaction.* https://www.testriq.com/blog/post/interoperability-testing-for-iot-ensuring-seamless-protocol-device-interaction

AIHA. (2016). *The future of sensors: Protecting worker health through sensor technologies.* American Industrial Hygiene Association.

AIHA. (2022). *Fact sheet: Reporting specifications for electronic real time gas and vapor detection equipment.* AIHA.

Aksut, G., Eren, T., & Alakas, H. M. (2024). Using wearable technological devices to improve workplace health and safety: An assessment on a sector base with multi-criteria decision-making methods. *Ain Shams Engineering Journal, 15*(2), 1–9.

Anderson, J. E. (2016). *Autonomous vehicle technology: A guide for policymakers.* Rand Corporation.

Andronie, M. L. (2023). Remote big data management tools, sensing and computing technologies, and visual perception and environment mapping algorithms in the internet of robotic things. *Electronics, 12*(1), 22. https://doi.org/10.3390/electronics12010022

Ayotte, D., & Corcoran, M. (2018). Individualized hyration plans improve perforamnce outcomes for collegiate athletes engaging in in-season training. *Journal of International Society of Sports Nutrition, 27,* 1–10.

Ballantine Jr., D., Martin, S., Ricco, A., Frye, G., Wohltjen, H., White, R., & Zellers, E. (1997). Chapter 5 - chemical and biological sensors. In *Acoutstic wave sensors: Theory, design, and physico-chemical applications* (pp. 222–330). Academic Press.

Bluetooth(R). (2024). *Understanding Bluetooth(R) range.* https://www.bluetooth.com/learn-about-bluetooth/key-attributes/range/#estimator

Bridgelall, R., & Tolliver, D. (2021). Budgeting for the adoption of sensors on connected trains. *Transportation Planning and Technology,* 1–17. https://doi.org/10.1080/03081060.2021.2017205

CA DOJ. (2024, March 13). *California consumer privacy act.* https://oag.ca.gov/privacy/ccpa

Cauda, M., Snawder, J., Spinazze, A., Cattaneo, A., Howard, J., & Cavallo, D. (2022). *The challenge for Industrial Hygiene 4.0: A NIOSH perspective on direct-reading methodologies and real-time monitoring in occupational environments.* CDC NIOSH.

Clark, J. (2018, December 13). Self-calibration and performance control of MEMS with applications for IoT. *Sensors, 18*(12). https://doi.org/10.3390/s18124411

Colasante, G., Makari, K., Hummel, T., & Murphy, C. (2024). Sample-to-answer direct real-time PCR detection of Anaplasma phagocytophilum, Ehrilchia spp, and Babesia spp. infections in whole-blook specimens. *Microbiol Spectr, 12,2,* 6–8.

Dhingra, L. S., et al. (2023). Use of wearable devices in individuals with or at risk for cardiovascular disease in the US, 2019 to 2020. *JAMA, 6*(6). https://doi.org/10.1001/jamanetworkopen.2023.16634

EEOC. (2000, July 27). *Enforcement guidance on disability - Related inquiries and medical examinations of employees under the ADA*. https://www.eeoc.gov/laws/guidance/enforcement-guidance-disability-related-inquiries-and-medical-examinations-employees

Elkateb, S., Métwalli, A., Shendy, A., & Abu-Elanien, A. E. (2024). Machine learning and IoT – Based predictive maintenance approach for industrial applications. *Alexandria Engineering Journal*, 298–309. https://doi.org/10.1016/j.aej.2023.12.065

European Commission. (2024, December 22). *Legal framework EU data protection*. European Commission. https://commission.europa.eu/law/law-topic/data-protection/legal-framework-eu-data-protection.en

Farnell. (2025). *Sensors for harsh industrial environments*. Farnell. https://fr.farnell.com/en-FR/technical-resources/articles/sensors-for-harsh-industrial-environments

Free, H., Collins, G., & Free, A. (1960). Triple-test strip for urinary glucose, Protein and pH. *Clinical Chemistry, 6*, 352–361.

Gander, P. (2023, November 2). The manufacturing evolution from Industry 1.0 to Industry 5.0. *Assured Partners*. Retrieved August 2024, from https://www.assuredpartners.com/news-insights/blogs/manufacturing/2023/the-manufacturing-evolution-from-industry-1-to-industry-5/

Gillis, A. S., & Yasar, K. (2025). What is IoT (internet of things)? *Ultimate IoT implementation guide for businesses*. https://www.techtarget.com/iotagenda/definition/Internet-of-Things-IoT

Government of Japan. (2023, April 1). *Laws and policies - act on the protection of personal information*. https://www.ppc.go.jp/en/legal

Gualandi, I., Tessarolo, M., Mariani, F., Possanzini, L., Scavetta, E., & Fraboni, B. (2021). Textile chemical sensors based on conductive polumers for the analysis of sweat. *Polymers, 6*. https://doi.org/10.3390/polym13060894

He, X. (2024, November 18). The future of sensor technology: Insights from IDTechEx research. *IDTechEx*. https://www.idtechex.com/en/research-article/the-future-of-sensor-technology-insights-from-idtechex-research/32109

Herrera, S., Petters, S., & Duffy, V. (2023). The incorporation of gamification into safety: A systematic review. In *HCI international – Late breaking papers* (Vol. 14058). https://doi.org/10.1007/978-3-031-48050-821

Huhn, S., Axt, M., Gunga, H., Maggioni, M., Munga, S., Obor, D., . . . Barteit, S. (2022, January 25). The impact of wearable technologies in health research: Scoping review. *JMIR mHealth and uHealth, 10*(1).

IEEE. (2020, October 23). *Hearing aid history: From ear trumpets to digital technology*. IEEE Pulse – Retrospectroscope. Retrieved March 2025, from https://www.embs.org/pulse/articles/hearing-aid-history-from-ear-trumpets-to-digital-technology/

ISC. (2019). *Electrochemical sensors cross interference table*. Industrial Scientific.

ISC. (2025, February 7). *Industrial scientific*. Retrieved February 7, 2025, from https://www.indsci.com/en/blog/common-gas-detector-sensor-types

ISO. (2023). *Accuracy (trueness and precision) of measurement methods and results — Part 1: General principles and definitions* (ISO 5725-1:2023). ISO.

ITU-T Y.2060. (2012). "Series Y: Global Information Infrastructure,Internet Protocol Aspects and Next-Generation Networks: Next Generation Networks – Frameworks and functional architecture models: Overview of Internet of Things"

Khalil, S., & Winkler, T. J. (2023). How software as a service simultaneously affords organizational agility and inertia. *The Journal of Strategic Information Systems.* https://doi.org/10.1016/j.jsis.2023.101804

Khan, I. (2016). Sensors are power hungry: An investigation of smartphone sensors' impact on battery power from a lifelogging perspective. *Bahria University Journal of ICT, 9,* 8–19. https://doi.org/10.1016/j.enss.2024.04.001

Leon, L. (2008, November 15). Thermoregulatory responses to environmental toxicants: The interaction of thermal stress and toxicant exposure. *Toxicology and Applied Pharmacology, 233*(1), 146–161.

Lippmann, M., & Gomez, M. R. (1996). Special report: Data elements for occupational exposure databases: Guidelines and recommendations for airborne hazards and noise. In *Applied, occupational and environmental hygiene* copyright © 1996 JOEH, LLC, reprinted by permission of Informa UK Limited, trading Taylor & Francis Group,https://www.tandfonline.com on behalf of JOEH, LLC..

Mamdiwar, S., R, A., Shakruwala, Z., Chadha, U., Srinivasan, K., & Chang, C. (2021). Recent advances on iot-assisted wearable sensor systems for healthcare monitoring. *Biosensors (Basel), 11*(10), 372. https://doi.org/10.3390/bios11100372

Marikyan, D. P. (2025). Technology acceptance model: A review. In S. Papagiannidis (Ed.), *TheoryHub Book.giannidis (Ed), TheoryHub Book.* Available at https://open .ncl.ac.uk / ISBN:9781739604400..

Masood, T., Sonntag, P. (2020). Industry 4.0:Adoption challenges and benefits for SMEs. *Computers in Industry,* 121,3. https://doi.org/10.1016/j.compind.2020 .103261.

Masson, J.-F. (2024). Roadmap for the use of machine learning and artificial intelligence in sensing. *ACS Sensors, 9*(8), 3805–3807. https://doi.org/10.1021/acssensors.4c02069

McKay, M. K., Lim, L. L., & Van Gelder, R. N. (2021). Rational laboratory testing in uveitis: A Bayesian analysis. *Survey of Ophthamology, 66*(5), 802–825.

Methlagl, M., Michimayer, F., & Perillo, V. (2023). Technological trust perceptions in wearable fitness technology: A person-centered approach. *Journal of Technology in Behavioral Science, 8,* 392–401.

Millar, C., Lockett, M., & Ladd, T. (2017, November). Disruption: Technology, innovation and society. *Technology Forecasting and Social Change, 129,* 2.

Mirion Technologies, INC. (2025). *Instadose VUE personal dosimeter.* Instadose. https://instadose.com/instadosevue

Mohammed, M., Khan, M. B., & Bashier, E. B. (2016). Introduction to machine learning. In M. B. Mohammed (Ed.), *Machine learning: Algorithms and applications* (pp. 1–34). CRC Press and Taylor & Francis https://doi.org/10.1201/9781315371658

Morris-Schaffer, K. A. (2021). A review of the LD50 and its current role in hazard communication. *ACS Chemical Health & Safety, 28*(1), 25–33.

Nagare, A. (2024). Ergonomics and usability in wearable tech: Balancing aesthetics, functionality, and comfort through industrial design. *International Journal of Innovative Research in Technology, 11*(6), 1482–1488.

National Safety Council (NSC). (2025). Work to Zero - Advancing Workplace Safety with Location Geofencing. 2–12.

NIOSH. (2024, August 12). *Direct reading sensors.* Centers for Disease Control and Prevention. https://www.cdc.gov/niosh/centers/direct-reading-sensor.html

NOAA / JPL. (2016). *National Environmental Satellite, Data, and Information Service (NESDIS).* National Oceanic and Atmospheric Administration.

OAIC. (2024, November 5). *Australian privacy principles*. https://www.oaic.gov.au/privacy/australian-privacy-principles.

Office of the Privacy Commissioner of Canada. (2025, September 9). *The personal informatoin protection and electronic documents act*. https://www.priv.gc.ca/en/privacy-topics/privacy-laws-in-canada/the-personal-information-protection-and-electronic-documents-act-pipeda/

OSHA. (1982, April 20). *1915.82 — Lighting*. U.S. Department of Labor Occupational Safety and Health Administration. https://www.osha.gov/laws-regs/regulations/standardnumber/1915/1915.82#:~:text=For%20landside%20areas%2C%20the%20employer%20shall%20provide,TO%20%C2%A7%201915.82%E2%80%94MINIMUM%20LIGHTING%20INTENSITIES%20IN%20FOOT%2DCANDLES.

Adarsh, R. (2025, December 23). *Industry 5.0 market report 2026: Key stats, figures & insights*. StartUs Insights. https://www.startus-insights.com/innovators-guide/industry-5-0-market-report/

Radanliev, P. (2025, June 17). Privacy, ethics, transparency, and accountability in AI systems for wearable devices. *Frontiers in Digital Health*. https://doi.org/10.3389/fdgth.2025.1431246

Robocraze. (2022, June 14). *Proximity sensor types*. Robocraze. https://robocraze.com/blogs/post/proximity-sensor-types#:~:text=They%20are%20used%20for%20detecting,and%20anti%2Dcollision%20safety%20systems.

Rogers, J. A. (2023). *Rogers research group*. Northwestern University. https://rogersgroup.northwestern.edu/

Rothfusz, L. (1990). *The heat index "equation" (or, more than you ever wanted to know about heat index) (NOAA Technical Attachment SR 90–23)*. National Oceanic and Atmospheric Administration (NOAA) Technical Attachment (SR 90–23).

Ruddro, R. A., & Mohana, H. A. (2023). Visual communication in industrial safety systems: A review of UI/UX design for risk alerts and warnings. *American Journal of Scholarly Research and Innovation, 2*(2), 217–245.

Sagarika, P., Mehlawat, S., Dharjwal, N., Kumar, A., & Sanger, A. (2024). Comprehensive review on gas sensors: Unveiling recent developments and addressing challenges. *Materials Science and Engineering, 308*, 117616.

Semtech. (2024). *Technology for a better world*. https://www.semtech.com/

Sevier, T. L., et al. (2000). The industrial athlete? *Work, 5*(3), 203–207.

Shockey, T., Dahm, M., Wurzelbacher & Baker, J. (2020, December). Industrial hygiene data standardization past lessons, present challenges, and future directions. *The Synergist*.

Siegel, D., Abrams, D., Hill, J., Jahn, S., Smith, P., & Thomas, K. (2019). *A practical guide for use of real-time detection systems for worker protection and compliance with occupational exposure limits*. AIHA.

Soga, K., & Schooling, J. (2016, August 6). Infrastructure sensing. *Interface Focus, 6*(4). https://doi.org/10.1098/rsfs.2016.0023

Spinelli, J. S. (2025). Wearable microfluidic biosensors with haptic feedback for continuous monitoring of hydration biomarkers in workers. *Nature, 8*(76). https://doi.org/10.1038/s41746-025-01466-9

StartUs Insights. (2025). *Sensor report 2025: Market data & tech Outlook*. StartUs Insights: https://www.startus-insights.com/innovators-guide/sensor-report/#Introduction

Stewart, M., Tineo, A., Woodrow, B., Wasik, M., & Chan, S. (2023). Continuous personal monitoring and personalized hydration recommendations with wearable sweat sensors to prevent occupational heat stress. In W. K. Ahram (Ed.), *Artificial intelligence, social computing and wearable technologies. 113.* AHFE International.

TEKTELIC Communications Inc. (2024, December 2). *LoRaWAN - most common applications and use cases.* https://www.iotforall.com/lorawan-most-common-applications-and-use-cases

The Law. (2023, November 10). *Privacy fundamental human right.* The Law Institute. Retrieved 1222,2024, from https://thelaw.institute/privacy-and-data-protection/privacy-fundamental-human-right/

Uhrovčík, J. (2014). Strategy for determination of LOD and LOQ values – Some basic aspects, Talanta. *Science Direct, 119,* 178–180.

Universal Technical Institute. (2024, August 24). *Robotics in manufacturing: Shaping the future of production.* https://www.uti.edu/blog/robotics-and-automation/robotics-in-manufacturing

US Department of Labor. (DOL) (n.d.). *Guidance on the protection of personally identifiable information (PII).* Retrieved December 22, 2024, from https://www.dol.gov/general/ppii

UT Health Houston. (2025). *Personal radiation dosimeter.* UTHealth Houston Radiation Safety. https://www.uth.edu/safety/radiation-safety/personal-radiation-dosimeter.htm

Vallée, A. (2024, March 5). Exoskeleton technology in nursing practice: Assessing effectiveness, usability, and impact on nurses' quality of work life, a narrative review. *BMC Nursing, 23*(1). https://doi.org/10.1186/s12912-024-01821-3

Wang, A., & Fu, L. (2024, November 22). Nano-functional materials for sensor applications. *Molecules, 29*(5515), 1–5.

Wang, B., Peng, J., Yang, K., Cheng, H., Yin, Y., & Wang, C. (2022, May 1). Multifunctional textile electronic with sensing, energy storing, and electrothermal heating capabilities. *ACS Appl Mater Interfaces.* 2022 May 18;14(19): 22497–22509. doi: 10.1021/acsami.2c06701. Epub 2022 May 6. PMID:35522598.

Welburn L, J. A. (2025). Prospects and trends in biomedical microelectromechanical systems (MEMS) devices: A review. *Biomolecules, 15*(6), 898. https://doi.org/10.3390/biom15060898

Wikipedia. (2023). *5G.* https://en.wikipedia.org/wiki/5G

Wikipedia. (2023). *IEEE 802.11.* https://en.wikipedia.org/wiki/IEEE_802.11

Wikipedia. (2024). *Bluetooth low energy.* https://en.wikipedia.org/wiki/Bluetooth_Low_Energy

Wikipedia. (2024). *LTE (Telecommunication).* https://en.wikipedia.org/wiki/LTE_(telecommunication)

Wikipedia. (2025, December 4). *Temperature.* https://en.wikipedia.org/wiki/Temperature

Windasari, N. A., & Lin, F. (2021). Why do people continue using fitness wearables? The effect of interactivity and gamification. *Sage Open, 11*(4). https://doi.org/10.1177/21582440211056606

Worksafe. (1994). *Safe working in a confined space* (National Standard NOHSC:1009(1994)). https://www.safeworkaustralia.gov.au/system/files/documents/1702/safeworkinginconfinedspaceaustralianstandards2865-1995nohsc1009-1994.pdf

Xin, P., Khan, F., & Ahmed, S. (2017). Dynamic hazard identification and scenario mapping using Bayesian network. *Process Safety and Environmental Protection,* 143–155. https://doi.org/10.1016/j.psep.2016.11.003

Zhang, C., Gorp, P., Derksen, M., Nuijten, R., Ijsselsteijn, W., Zanutto, A., . . . Pratola, R. (2021). Promoting occupational health through gamification and E-Coaching: A 5-month user engagement study. *International Journal of Environmental Research and Public Health*. https://doi.org/10.3390/ijerph18062823

Zhao, X., Lin, C., Wang, J., & Oh, D. (2014). Advances in rapid detection methods for foodborne pathogens. Microbiology and Biotechnology, *24*, 297–312.

Zhao, Y. W. (2025). A review of wearable flexible sensors for sports: From materials to applications. *International Journal of Smart and Nano Materials, 16*(4), 695–725. https://doi.org/10.1080/19475411.2025.2552803

ZigBee Alliance. (2013). *ZigBee specification FAQ*. web.archive.org/web/20130627172453/http://www.zigbee.org/Specifications/ZigBee/FAQ.aspx

Index

I

J

L

M

N

O

P

R

S

T

U

V

W

For Product Safety Concerns and Information please contact our EU representative GPSR@taylorandfrancis.com
Taylor & Francis Verlag GmbH, Kaufingerstraße 24, 80331 München, Germany

www.ingramcontent.com/pod-product-compliance
Ingram Content Group UK Ltd.
Pitfield, Milton Keynes, MK11 3LW, UK
UKHW020208200926
475702UK00004B/669
9781032914800